Importance of Selenium in the Environment and Human Health

Edited by Mohammed Muzibur Rahman,
Abdullah Mohamed Asiri,
Anish Khan and Inamuddin

Published in London, United Kingdom

IntechOpen

Supporting open minds since 2005

Importance of Selenium in the Environment and Human Health
http://dx.doi.org/10.5772/intechopen.83202
Edited by Mohammed Muzibur Rahman, Abdullah Mohamed Asiri, Anish Khan and Inamuddin

Contributors
Aleksandra Sentkowska, Mohammed Muzibur Rahman, Olga Senkevich, Yuriy Koval'Skiy, Letícia De Abreu Faria, Pedro Henrique De Cerqueira Luz, Adibe Luiz Abdala, Yibin Feng, Ning Wang, Chien-Shan Cheng, Abiodun Humphrey Adebayo, Omolara Faith Yakubu, Osarobo Bakare-Akpata

First published in London, United Kingdom, 2020 by IntechOpen
IntechOpen is the global imprint of INTECHOPEN LIMITED, registered in England and Wales, registration number: 11086078, 7th floor, 10 Lower Thames Street, London, EC3R 6AF, United Kingdom
Printed in Croatia

British Library Cataloguing-in-Publication Data
A catalogue record for this book is available from the British Library

Additional hard and PDF copies can be obtained from orders@intechopen.com

Importance of Selenium in the Environment and Human Health
Edited by Mohammed Muzibur Rahman, Abdullah Mohamed Asiri, Anish Khan and Inamuddin
p. cm.
Print ISBN 978-1-78985-867-9
Online ISBN 978-1-78985-868-6
eBook (PDF) ISBN 978-1-78984-312-5

Meet the editors

Mohammed Muzibur Rahman received his BSc and MSc from Shahjalal University of Science and Technology, Sylhet, Bangladesh, in 1999 and 2001, respectively. He received his PhD from the Chonbuk National University, South Korea, in 2007. After his PhD, he worked as a postdoctoral fellow and assistant professor in pioneering research centers and universities located in South Korea, Japan, and Saudi Arabia (2007–2011). Since 2011 he is working as an associate professor in the Center of Excellence for Advanced Materials Research (CEAMR) and the Chemistry Department at King Abdulaziz University, Saudi Arabia. He has published papers in more than 290 international and domestic conferences, as well as several book chapters and 10 books as an editor. His research work has been largely in the area of photocatalysis, semiconductors, nanoparticles, carbon nanotubes, nanotechnology, electrocatalysis, sensors, ionic liquids, surface chemistry, gold nanoparticles, electrochemistry, instrumental science, nanomaterials, self-assembled monolayers, photochemistry, μ-chips, and devices.

Abdullah Mohamed Asiri received his PhD from the University of Wales, College of Cardiff, UK, in 1995. He has been the Head of the Chemistry Department at King Abdulaziz University since October 2009 and is the founder and director of the Center of Excellence for Advanced Materials Research (CEAMR). He is also a professor of organic photochemistry. His research interest covers color chemistry, synthesis of novel photochromic and thermochromic systems, synthesis of novel coloring matters and dyeing of textiles, materials chemistry, nanochemistry, nanotechnology, polymers, and plastics. He is the editor-in-chief of the King Abdul Aziz University Journal of Science. He is also a member of the editorial board of Pigments and Resin Technology (UK), Organic Chemistry in Sight (New Zealand), and Recent Patents on Materials Science (USA). He is the Vice-President of the Saudi Chemical Society (Western Province Branch).

Dr. Anish Khan is currently working as an assistant professor at the Chemistry Department, Centre of Excellence for Advanced Materials Research (CEAMR), Faculty of Science, King Abdulaziz University, Jeddah, Saudi Arabia. He received his PhD from Aligarh Muslim University, India, in 2010. He has research experience of working in the field of synthetic polymers and organic–inorganic electrically conducting nanocomposites. He completed his postdoctoral studies in electroanalytical chemistry from the School of Chemical Sciences, University Sains Malaysia (USM), in 2010–2011. He has published more than 127 research papers in refereed international journals and more than 20 papers in international conferences/workshops, as well as five books, 14 book chapters, and 14 chapters in progress. His research includes the preparation and characterization of hybrid nanocomposite materials and their applications, polymeric inorganic cation exchange materials, electrically conducting polymeric materials, composite material use as sensors, green chemistry by remediation of pollution, heavy metal ion selective membrane electrodes, and biosensors on neurotransmitters.

Dr. Inamuddin is currently working as an assistant professor in the Chemistry Department, Faculty of Science, King Abdulaziz University, Jeddah, Saudi Arabia. He has extensive research experience in multidisciplinary fields of analytical chemistry, materials chemistry, electrochemistry, and more specifically, renewable energy and the environment. He has published 127 research articles in international journals of repute and 18 book chapters in knowledge-based book editions published by renowned international publishers. He has published 39 edited books with Springer, United Kingdom, Elsevier, Nova Science Publishers, Inc. USA, CRC Press Taylor & Francis, Asia Pacific, Trans Tech Publications Ltd., Switzerland, and Materials Science Forum, USA. He is a member of various editorial boards serving as associate editor for journals such as Environmental Chemistry Letter, Applied Water Science, Euro-Mediterranean Journal for Environmental Integration, Springer-Nature, Scientific Reports-Nature, and the editor of Eurasian Journal of Analytical Chemistry.

Contents

Preface

Importance of Selenium in the Environment and Human Health contains a number of recent studies on the analyses of the effect of selenium on environmental as well as human health. Selenium is very significant in environments, animals, plants, microorganisms, and humans. It is a beneficial microelement existing in minute amounts in environments as well as organisms. It is also very useful to animals as well as human health as an active antioxidant, hormonal regulator, and anticarcinogenic. However, it can be toxic at excess levels of concentration, which results in competing and replacing sulfur in amino acids. Excess levels lead to inappropriate folding of protein and eventually the creation of a non-functional protein and enzymes in living beings. Selenium is also a beneficial micronutrient, which reduces the risk of cancer in human health. As an essential trace element, it has been studied for its anticancer properties in both oxidative stress and in inflammatory-related mechanisms that may contribute to hepatocellular carcinoma growth and metastasis. This book looks at the intraconnections and interrelationships between selenium in the environment, plants, agriculture, biology, human health, animals, and molecular and biochemistry processes. It is a key book for research organizations, governmental research centers, academic libraries, and R&D facilities affiliated to the recent research and study of the effects of selenium on environmental and human health.

Prof. Mohammed Muzibur Rahman, Prof. Abdullah Mohamed Asiri, Dr. Anish Khan and Dr. Inamuddin
Center of Excellence for Advanced Materials Research (CEAMR),
Department of Chemistry,
Faculty of Science,
King Abdulaziz University,
Jeddah, Saudi Arabia

Chapter 1

Introductory Chapter: Fundamental Discussion of Selenium Effect

Mohammed Muzibur Rahman

1. Introduction

Here, it contains a number of recent researches on the study of selenium as well as their effect in environment and human health. Generally, in environmental samples selenium can exist in inorganic forms and as organic species with direct carbon-selenium bonds. Thus, the development of reliable techniques to study the speciation and isolation of selenium in environmental samples is necessary. Beside that, selenium is existed into minerals of fundamental importance for human health as well as living plants. Selenium status in general population is very important due to its remarkable benefits to the human body as antioxidant, hormonal regulator, and anticarcinogenic. On the other way, selenium can be toxic leaving representing a multitude of disciplines from academic, industry, and governments for sharing their extraordinary innovative and unique knowledge on this Selenium results. Moreover, selenium is perhaps the naturally occurring trace rare-earth metallic element of extreme concern worldwide. The excess amounts of selenium can lead to toxicosis and teratogenesis in plants and animals, while the permanent effect of selenium insufficiency can be even more noteworthy. Both excess and deficit are known to cause a wide range of clinical manifestations. Even though a large body of evidence provides the information about selenium, the exact molecular mechanisms of its effect in physiologic and pathologic conditions remain unknown.

2. Literature survey

Generally, selenium status in general population is very important due to its remarkable benefits to the human body as antioxidant, hormonal regulator, and anticarcinogenic. However, the relationship between selenium and health has been the focus of medical community. Early observational studies have shown that the trace element selenium in the environment is closely related to the occurrence and development of tumors [1]. The incidence of cancer in patients with selenium deficiency is significantly increased, and the amount of selenium in the body is negatively correlated with cancer [2]. Furthermore, while some studies suggested selenium supplementation reduce the risk of cancer, some methodologically sound trials suggested selenium supplementation does not reduce the risk of cancer and may even increase it for some types, including advanced prostate cancer and skin cancer [3, 4]. Cattle in Brazil are fed basically from pasture, but there are strong evidences that soils contain low availability of Se; consequently plants and animals incorporate low Se levels. Pastures Se fertilized bring benefits to nutrition and

IntechOpen

health of animal and consequently humans already known in some countries. In contrast, Se fertilization on tropical weathered soils and tropical forages are little known. Selenium exists in many different inorganic materials. Inorganic selenium, present in water and soil, can be easily transformed into volatile compounds by plants and fungi. Organic species of selenium form covalent C-Se bonds. SeCys is included into selenoproteins and participate in redox reactions. The metabolic pathway of selenium in human body is complicated [5]. In young healthy non-pregnant women, the existence of a direct relationship of moderate strength between selenium content in blood plasma and urine is established. There is a highly significant direct strong correlation between selenium in serum and erythrocytes and medium-strength relationship between selenium in urine and selenium in erythrocytes. At the same time, during the physiological course of the gestational process in healthy mothers, serum selenium levels and urine selenium values have a direct relationship of average strength. For the growth and development of the fetus, normal pregnancy requires a constant consumption of sufficient amounts of nutrients; at least 40 of them are essential for pregnant and lactating women. Nutritional status of pregnant and lactating women is of great importance for the health of the child, as the development of the body is most actively carried out up to 18 months of life. Some earlier surveying studies found that low environmental selenium is associated with certain cancers in the digestive system and selenium supplementation may provide some cancer prevention effect [6].

Moreover, the agronomic biofortification of food through field fertilization with Se could be a solution to provide this micronutrient for animals and humans through plants. The plants are able to absorb and incorporate Se to organic compounds as seleno-amino acids. Thus, inorganic Se is converted to organic Se compounds through the plants which can be easily absorbed by the human body and be available where needed in the body [7]. In some countries, Se fertilization is well established, and it is annually made in New Zealand in which Se is applied along with phosphorus fertilization in pastures [8]. The interference of large amounts of chloride ions during selenium atomization was prevented by using iridium as a permanent chemical modifier. Clinical and diagnostic parallels were made to assess the role of trace element imbalance in the placenta in the formation of pathological conditions of newborns in the early neonatal period by calculating the relative risk with 95% confidence interval. The low natural levels of Se in soils and its absence in fertilization to crops explain the low contents in food from vegetables and consequently in Brazilian diet, except for North areas. However, in the study of hemodialysis patients from North and Southeast of Brazil, both patient groups presented low Se in plasma levels, when compared to recommend standard values; independent of the region, all patients are found the Se deficiency [9]. Concisely, the concentration of selenium in serum is reduced in children with intraventricular hemorrhage and diseases of the gastrointestinal tract. However, first of all, premature children are a risk group for selenium deficiency, aggravated by living in environmentally unfavorable conditions. Many of them, especially those receiving long-term inpatient treatment, are artificially fed, and baby foods contain predominantly sodium selenite—an inorganic compound of selenium with high toxicity and low bioavailability and not always in sufficient quantities. Selenium obtained from mother's milk is better absorbed than selenium nutrient mixtures. It is recommended to add selenium to the nutrition of mothers, as well as cows whose milk is used for the preparation of nutrient mixtures. Establishment of effective and safe rates in tropical environmental is still required and which is unknown regardless of the implication to the plant enrichment of Se. High Se availability in agriculture soils can cause toxicity to crops, but it is still more concerning if a crop shows accumulator character, i.e., if a crop has capacity of absorb high levels of selenium with no symptoms of toxicity, this could increase the possibilities to cause toxicity for animals or human.

In this study, cation-exchange chromatography was used to analyze selenium-enriched yeast in a human adsorption study [10]. A mobile phase pyridinum formate buffer with 3% of methanol was used. This method was suitable for the separation of organic selenium species, however not suitable to separate the selenite. There were no differences in the content of selenium in the serum of young healthy nonpregnant women, donor volunteers, healthy pregnant women at the end of physiological pregnancy, and healthy puerperas. The provision in these groups was found to be average (82–85 μg/l), which is approximately 80% of the optimal level, since the interval of normal serum concentrations of selenium is on average 115 μg/l. At the end of physiological pregnancy and normal childbirth, the level of selenium in the blood serum was determined, which is lower than optimal for pregnant women. In premature births, serum levels of selenium were significantly lower, which is about half of the optimum. There are no significant differences in the content of selenium in the blood inversion in smoking and non-smoking women in the indicator group or in the group of healthy pregnant women. But there are significant differences found in the hair of smoking and non-smoking pregnant women. Hill et al. who investigated human liver cancer HepG2 cell line showed that in selenium-deprived HepG2 cells, selenoprotein P release decreased to 10% [11]. Further, various studies consistently reported apoptosis induction effect of selenite in human hepatoma cells HepG2 cells, potentially by inducing the release of lactate dehydrogenase (LDH) and decreasing glutathione (GSH) production [12–14]. Another study reported that selenite-induced apoptosis in HepG2 cells was mediated by reactive oxygen species (ROS) that activated JNK to regulate apoptosis [15]. A more recent study on selenium nanoparticles surface decorated with Galangin can induce apoptosis through p38 and AKT signaling pathway in HepG2 cells [16]. Similarly, selenium nanoparticles synthesized with extract of hawthorn fruit also induced apoptosis in HepG2 cells [17].

High Se contents in leaves from tropical grass and in legumes comprise another fact to be analyzed for Se fertilization rate establishment despite the few data. Usually, grass leaves and legumes are preference fractions for cattle according its intake selective behavior, mainly in tropical pastures due to high accumulation of stem portions and its low digestibility. Selenium is one of the few elements absorbed by plants in enough quantities able to intoxicate domestic animals [18]. Selenium and selenium supplementation for the treatment of liver disease should attract the attention of the medical community. However, controversies remain with whether a relationship exists between serum selenium level and HCC risk. In the study of selenium in breast milk of women living in the United States, it was determined that the average selenium content is 18 μg/l, and the maximum level reached 60 μg/l; a direct correlation was found between the level of selenium in milk and serum. In premature births, the milk of women living in New Zealand contains an average of 20 μg/l selenium. The question of the needs of newborns of selenium is not finally resolved, but most researchers recommend enriching the mixture for children with selenium in an amount corresponding to its content in breast milk. But the supply of children with selenium in an amount equal to its content in breast milk is not equivalent, because breast milk and mixtures contain different chemical forms of selenium with different levels of bioavailability and toxicity. However, epidemiological investigations and biological studies should be further conducted to demonstrate and verify whether selenium supplements are beneficial for the prevention and treatment of HCC and to elucidate its exact mechanism of action. Obtaining separation was satisfied for organic compounds but poor for selenite and selenite. For separation of organic and inorganic forms of selenium, tetrabutylammonium acetate was proposed [19]. Also mixed ion-pairing reagents (butanesulphonic and tetramethylammonium hydroxide) were also used to simultaneously separate inorganic and organic species with satisfactory separation efficiencies [20]. The most pronounced imbalance of trace elements in the groups of premature infants with a gestation period of 28–33 weeks and in newborns with intrauterine growth retardation syndrome is found, which is a manifestation of immaturity in the first and decompensation of the function of passive transport systems in the second case. Clinically, this will be manifested by a violation of acute neonatal adaptation with a low Apgar score, severe general condition, and the presence of markers of inflammatory response.

In the epidemiological study of the provision of selenium in children permanently living in Khabarovsk, a significant variability in the provision of selenium in different age groups of children with maximum security in adolescents 12–17 years is established, and the minimum level of selenium was detected in children aged 2 years of life, due to the peculiarities of nutrition and food preferences of children of different ages. In general, selenium deficiency was observed in 18% of the surveyed children, and only 28% of children found trace element content at the lower limit of the norm. The data obtained by us on the provision of selenium for children and residents of Khabarovsk are comparable with the data obtained by us from the adult population. In serum and placenta, selenium to some extent mimics the behavior of zinc and behaves opposite to copper. It is interesting to note that active smoking gave higher levels of selenium and zinc in the placenta. Both elements play an important role in protecting against oxidant stress: selenium in glutathione peroxidase and zinc in superoxide dismutase. Smoking causes severe chemical stress to the placenta, so a higher concentration of zinc and selenium in the placenta in smokers may reflect the activation of protective mechanisms.

3. Conclusion

Finally, selenium can be toxic leaving representing a multitude of disciplines from academic, industry, and governments for sharing their extraordinary innovative and unique knowledge on this Selenium results. Moreover, selenium is perhaps the naturally occurring trace rare-earth metallic element of extreme concern worldwide. The excess amounts of selenium can lead to toxicosis and teratogenesis in plants and animals, while the permanent effect of selenium insufficiency can be even more noteworthy. Both excess and deficit are known to cause a wide range of clinical manifestations. Even though a large body of evidence provides the information about selenium, the exact molecular mechanisms of its effect in physiologic and pathologic conditions remain unknown. This book explored the connection and interrelationships between selenium in environment, plants, agriculture, biology, human health, animals, molecular, and biochemistry process to complete this book. It is an important booklet for research organizations, governmental research centers, academic libraries, and R&D affianced in recent research and studied selenium effect in the environmental and human health.

Author details

Mohammed Muzibur Rahman
Department of Chemistry, Faculty of Science, King Abdulaziz University, Jeddah, Saudi Arabia

*Address all correspondence to: mmrahmanh@gmail.com

References

[1] Vinceti M, Filippini T, Cilloni S, et al. Health risk assessment of environmental selenium: Emerging evidence and challenges (Review). Molecular Medicine Reports. 2017;**15**(5):3323-3335

[2] Zhang Z, Bi M, Liu Q, Yang J, Xu S. Meta-analysis of the correlation between selenium and incidence of hepatocellular carcinoma. Oncotarget. 2016;**7**(47):77110-77116

[3] Vinceti M, Filippini T, Del Giovane C, et al. Selenium for preventing cancer. Cochrane Database of Systematic Reviews. 2018;**1**:CD005195

[4] Vinceti M, Filippini T, Cilloni S, Crespi CM. The epidemiology of selenium and human cancer. Advances in Cancer Research. 2017;**136**:1-48

[5] Wastney ME, Combs GF, Canfield WK, Taylor PR, Patterson KY, Hill DA, et al. A human model of selenium that integrates metabolism from selenite and selenomethionine. Journal of Nutrition. 2011;**141**:708-717

[6] Shenkin A. Selenium in intravenous nutrition. Gastroenterology. 2009;**137** (Suppl 5):S61-S69

[7] Agency for Toxic Substances and Disease Registry (ATSDR). Toxicological Profile for Selenium. Atlanta, GA: U.S. Department of Health and Human Services, Public Health Service; 2003

[8] Watkinson JH. Prevention of selenium deficiency in grazing animals by annual top dressing of pasture with sodium selenate. New Zealand Veterinary Journal. 1983;**31**:78-85

[9] Stockler-Pinto MB, Malm O, Azevedo SRG, Farage NE, Dorneles PR, Cozzolino SMF, et al. Selenium plasma levels in hemodialysis patients: Comparison between north and southeast of Brazil. Jornal Brasileiro de Nefrologia. 2014;**36**(4):490-495

[10] Larsen EH, Sloth J, Hansen M, Moesgaard S. Selenium speciation and isotope composition in 77Se enriched yeast using gradient elution HPLC separation and ICP-dynamic reaction cell-MS. Journal of Analytical Atomic Spectrometry. 2003;**18**:310-316

[11] Hill KE, Chittum HS, Lyons PR, Boeglin ME, Burk RF. Effect of selenium on selenoprotein P expression in cultured liver cells. Biochimica et Biophysica Acta. 1996;**1313**(1):29-34

[12] Shen H, Yang C, Liu J, Ong C. Dual role of glutathione in selenite-induced oxidative stress and apoptosis in human hepatoma cells. Free Radical Biology & Medicine. 2000;**28**(7):1115-1124

[13] Shen HM, Ding WX, Ong CN. Intracellular glutathione is a cofactor in methylseleninic acid-induced apoptotic cell death of human hepatoma HEPG(2) cells. Free Radical Biology & Medicine. 2002;**33**(4):552-561

[14] Celik HA, Aydin HH, Deveci R, et al. Biochemical and morphological characteristics of selenite-induced apoptosis in human hepatoma Hep G2 cells. Biological Trace Element Research. 2004;**99**(1-3):27-40

[15] Zou Y, Niu P, Yang J, Yuan J, Wu T, Chen X. The JNK signaling pathway is involved in sodium-selenite-induced apoptosis mediated by reactive oxygen in HepG2 cells. Cancer Biology & Therapy. 2008;**7**(5):689-696

[16] Li Y, Guo M, Lin Z, et al. Multifunctional selenium nanoparticles with galangin-induced HepG2 cell apoptosis through p38 and AKT signalling pathway. Royal Society Open Science. 2018;**5**(11):180509

[17] Cui D, Liang T, Sun L, et al. Green synthesis of selenium nanoparticles with extract of hawthorn fruit induced HepG2 cells apoptosis. Pharmaceutical Biology. 2018;**56**(1):528-534

[18] Dhillon KS, Dhillon SK. Distribution and management of seleniferous soils. Advances in Agronomy. 2003;**79**:119-184

[19] Marchante-Gayon JM, Thomas C, Feldmann I, Jakubowski N. Comparison of different nebulizers and chromatographic techniques for the separation of selenium in nutritional commercial supplements by hexapole collision and reaction cell ICP-MS. Journal of Analytical Atomic Spectrometry. 2000;**15**:1093-1102

[20] Zheng J, Kosmus W. Retention study of inorganic and organic selenium compounds on a silica-based reversed phase column with ion-pairing reagents. Chromatographia. **51**:338-344

Chapter 2

The Importance of Selenium in Children's Health and Reproduction

O.A. Senkevich and Y.G. Koval'skiy

Abstract

The authors report selenium deficiency in pregnant women, which can lead to an increase in the frequency and severity of early and late gestosis, fetal hypotrophy, hypoxia, and increased risk of miscarriage. The provision of selenium in children depends on the degree of maturity and physical development, decreases with fetal hypotrophy, prematurity, artificial feeding, and hypoxia. The content of selenium in breast milk of women who gave birth prematurely, contains selenium three times less than in normal childbirth, which contributes to a high risk of alimentary-dependent conditions in premature infants.

Keywords: selenium, children, pregnant women, reproduction, placenta, breast milk

1. Introduction

Pregnancy is a period of important conditional physiological changes, when the fetus requires a regular and balanced diet provided by the mother's food and its physiological reserves. The intrauterine period, infancy, and early age are considered to be a critical period in terms of the impact of nutrition on subsequent development [1].

For the growth and development of the fetus, normal pregnancy requires a constant consumption of sufficient amounts of nutrients, at least 40 of them are essential for pregnant and lactating women. Nutritional status of pregnant and lactating women is of great importance for the health of the child, as the development of the body is most actively carried out up to 18 months of life.

2. Value of selenium deficiency in the perinatal period

Adequate provision of selenium for pregnant women, premature infants, children of different ages, and adolescents living in environmentally unfavorable conditions and constituting a risk group for selenium deficiency is particularly relevant [2–7].

Nefedova et al. [8] believe that the fact of selenodetic in healthy women in Western Siberia is associated with the possible formation of "anomalous biogeochemical province" endemic selenium. In pregnancy, even with the physiological course, this deficit is exacerbated, which is an understandable increased expenditure of trace elements [9–11]. In women at risk of termination of pregnancy, selenium deficiency is most pronounced, with pregnancy ending in miscarriage at different stages [12–15].

The level of selenium (serum, erythrocytes, urine, and hair) in healthy adult blood donors, residents of Khabarovsk in the preconception (indicator group), at the end of physiological pregnancy (group "healthy pregnant") and breast milk on the 7–10 day after birth in mothers of healthy newborns and preterm infants was established.

There were no differences in the content of selenium in the serum of young healthy nonpregnant women, donor volunteers (residents of Khabarovsk, examined by random sampling), healthy pregnant women at the end of physiological pregnancy, and healthy puerperas. The provision in these groups was found to be average (82–85 μg/l), which is approximately 80% of the optimal level, since the interval of normal serum concentrations of selenium is on average 115 μg/l. At the end of physiological pregnancy and normal childbirth, the level of selenium in the blood serum was determined within 85.4 + 4.8 μg/l and 82.6 + 6.1, respectively, which is lower than optimal for pregnant women. In premature births, serum levels of selenium were significantly lower—58.2 μg/l, which is about half of the optimum. There are no significant differences in the content of selenium in the blood inversion in smoking and nonsmoking women in the indicator group (82.8 + 16.2 and 83.1 + 15.9 μg/l) or in the group of healthy pregnant women (84.7 + 15.8 and 85.8 + 16.1 μg/l), in the presence of significant differences in the hair of smoking and nonsmoking pregnant women (383.0 + 24.4 and 436.6 + 28.2 μg/l, respectively, $p < 0.01$).

During pregnancy, there is a significant reduction in the excretion of selenium in the urine, which ensures the maximum possible provision of the fetus with a trace element [16]. Excretion of selenium in urine is constant and is normally 40–50% of intake [17]. Losses of selenium with urine in the group of healthy pregnant women in Khabarovsk ranged from 12.6 to 40.8 μg/l, on average 23, 68 μg/l, which differs with a high degree of reliability from the standards described in the literature.

A direct relationship of average strength between the level of selenium in the hair and urine of pregnant women indicate unidirectional changes in the concentration of selenium in the hair and urine of pregnant women.

Also in the process of gestation, there is a change in the content of selenium in serum and in hair, and there is a highly reliable feedback of average strength between the content of selenium in serum and hair of pregnant women.

In young healthy nonpregnant women, the existence of a direct relationship of moderate strength ($r = 0.4804$) is established between selenium content in blood plasma and urine, and there is a highly significant direct strong correlation between selenium in serum and erythrocytes ($r = 0.9552$), direct medium strength correlation between selenium in urine and selenium in erythrocytes ($r = 0.4348$). At the same time, during the physiological course of the gestational process in healthy mothers, serum selenium levels and urine selenium values have a direct relationship of average strength.

The loss of selenium in urine was significantly lower ($p < 0.05$) in women with a long labor period (18.6 + 0.76 μg/l) and higher in mothers with planned operative labor (24.9 + 1.3 μg/l), and there was no dependence on the gestation period. Probably, the detected changes are associated with the development of oxidant stress; since the intensive physical activity (childbirth) determines the acceleration of metabolic processes, leading to significant oxidant stress of the body, a specific mechanism is included that ensures the preservation of selenium by reducing its excretion in the urine [18].

On the territory of the Khabarovsk, there is a shortage of selenium, because in the food diet of the inhabitants of the region, local products prevail, including from household plots, which, in conditions of natural low maintenance, contain little of this element. The problem is also aggravated by the fact that our study of the diet of young children showed a significant deficit in the consumption of the surveyed residents of Khabarovsk products—sources of selenium, which undoubtedly exacerbates the population deficit.

In the epidemiological study of the provision of selenium in children permanently living in Khabarovsk, we have established a significant variability in the provision of selenium in different age groups of children with maximum security in adolescents 12–17 years; the minimum level of selenium was detected in children aged 2 years of life, due to the peculiarities of nutrition and food preferences of children of different ages. In general, selenium deficiency was observed in 18% of the surveyed children; only 28% of children found trace element content at the lower limit of the norm. The data obtained by us on the provision of selenium for children and residents of Khabarovsk are comparable with the data obtained by us from the adult population [19].

3. Placenta as an indicator of fetal selenium content

The placenta is a unique, complex organ composed of the maternal and fetal parts. It is a full component of the system and performs numerous functions, the violation of which leads to a wide range of pregnancy complications [20]. The placenta has a wide range of functions that go in both ways; directly from maternal body to placenta; from placenta to fetus, and vice versa, from the fetus to the mother and placenta [21–23]. Placental dysfunction is a threat to the development and life of the fetus, and subsequently the newborn. In physiological pregnancy, the mother's body retains its homeostasis and provides the fetus with everything necessary for normal development [24, 25]. Studies in recent years [3, 26] show that the inability of the system of "mother-placenta-fetus" to maintain adequate exchanges between mother and fetus (feto-placental insufficiency) leads to violation of fetal development and homeostasis of the mother. The value of the placenta is as an indicator of fetal selenium content [20–22, 27].

It is known that the mass of the newborn is largely determined by serum selenium [21, 28]; however, the unknown is the correlation between the content of selenium level with the mass of the placenta and fetus. During the multivariate correlation analysis, it was found that there is an average strength positive correlation ($R = 0.53$; $p = 0.002$) between the body weight of premature infants born at gestation 28–36 weeks, and the content of selenium in the placenta.

Therefore, the higher the level of selenium in the placenta, the greater the body weight of the newborn. The same relationship was established in the group of children with psrp ($R = 0.57$; $p = 0.03$). The obtained results do not confirm the data of Zadrozna et al. [27], which did not reveal this dependence. A negative correlation between the placental-fetal coefficient and the content of selenium in the placenta ($R = -0.55$; $p = 0.002$) was established [29].

In the study of the accumulation of selenium in the placenta, depending on the duration of pregnancy, it was found that the minimum content was observed at the gestation period of 28–33 weeks and significantly different from the indicators in other groups [29]. The content of selenium in the placenta of women in countries such as Russia, Ukraine, Poland, Spain, and Turkey is in the range of 0.15–1.65 mg/kg [21–23, 30, 31], which is significantly higher.

There is a positive correlation between the average strength between the levels of selenium in the placenta and the evaluation of the newborn on the Apgar scale both at the first ($R = 0.45$; $p = 0.02$) and fifth minute ($R = 0.36$; $p = 0.005$), which indicates a better tolerability of labor stress in newborns with a greater reserve of this microelements antioxidant properties in the placenta.

Clinical and diagnostic parallels were made to assess the role of trace element imbalance in the placenta in the formation of pathological conditions of newborns in the early neonatal period by calculating the relative risk (RR) with 95% confidence interval.

It was found that the deficiency of selenium significantly and statistically significantly affects the low score on the Apgar scale in the first minute in newborns—more often 2.5 times (OR with CI 1.1–2.5) for selenium. A low content of selenium in the placenta in newborns significantly increases the risk of neutrophilia with OR—1.6 (CI 1.1–3.0). The above parallels occur reliably in all study groups. In the pathogenesis of RDS, selenium deficiency probably plays a significant role because its deficiency not only in the hair of the mother and newborn, but also in the placenta at gestation 28–33 weeks forms the risk of RDS and is 1.9 CI 1.06–4.3.

It was found that the most pronounced imbalance of trace elements in the groups of premature infants with a gestation period of 28–33 weeks and in newborns with intrauterine growth retardation syndrome, which is a manifestation of immaturity in the first and decompensation of the function of passive transport systems in the second case. Clinically, this will be manifested by a violation of acute neonatal adaptation with a low Apgar score, severe general condition, and the presence of markers of inflammatory response.

In serum and placenta, selenium to some extent mimics the behavior of zinc and behaves opposite to copper [32]. It is interesting to note that active smoking gave higher levels of selenium and zinc in the placenta [32]. Both elements play an important role in protecting against oxidant stress: selenium in glutathione peroxidase and zinc in superoxide dismutase. Smoking causes severe chemical stress to the placenta, so a higher concentration of zinc and selenium in the placenta in smokers may reflect the activation of protective mechanisms.

When smoking, the transfer of selenium from the blood to the placenta is increased, so that the level of selenium in the blood decreases; while the level of selenium in the placenta of smoking women is much higher than in nonsmokers. In general, the results suggest that in women smokers, selenium transport appears to be part of a protective mechanism against chemical stress.

Selenium concentrations during pregnancy are reduced in serum and placenta by the end of pregnancy and increased in the placenta of smokers. Since selenium status negatively affects the level of zinc in breast milk [33, 34], the concentration of selenium can affect the level of zinc in the placenta and serum by changing the distribution of zinc. Low concentrations of serum selenium and low copper concentrations in the placenta are associated with higher weight of newborn, which reflects the correlation between serum selenium and copper of the placenta. A strong positive correlation in serum between selenium and copper correspond to such in milk [34], although the reason for this is not clear. An inverse correlation between the activity of aromatase and placenta selenium was revealed, which indicates the protective effect of selenium by acting on the activity of the enzyme to transfer nutrients to the fruit. The authors conclude that it is essential to identify the relationship between selenium, zinc, and copper. Thus, the protective role is established selenium during pregnancy, especially in smokers (**Table 1**).

It is shown that Cd and Pb placenta have an inverse correlation with the body weight of the newborn [25, 36], and these elements are selenium antagonists which reduce their toxic effect [36–38].

In Croatia, Alexiou et al., [35], indicated that Se in the placenta predicts birth of a child of normal weight, not intrauterine growth restriction. In this work, a significant positive relationship between placental Se and neonatal weight in a

Indicator	Level	Region
Se placentae	810 ± 20 μg/kg	Spain*
Se umbilical cord blood	74 ± 7 μg/kg (51–104)	
Se mothers blood	90 ± 15 (57–118)	
Se mothers hair	600 ± 370 (220–1500)	
Se newborn hair	1040 ± 480 (400–2530 μg/kg)	
Placentae	200 μg/kg raw weight	Austria**
Placentae		Italy***
Placentae (normal pregnancy) (36)	150 ± 30 (100–240 мкг/кг)	Croatia**** (significant differences with the norm $P < 0.05$)
Placentae intrauterine growth restriction cases (49)	140 ± 20 (100–200 мкг/кг)	

Lorenzo Alonso et al. [22, 23].
**Iyengar and Raphuman [31].*
***Capellia et al. [21].*
****Alexiou et al. [35].*

Table 1.
Indicators of selenium in the placenta.

subgroup of pregnant women with body weight appropriate for this gestational age was established. This is consistent with a number of important functions that can be used to influence fetal development. On the other hand, the lack of communication in the IWRM group may mean that the element is not causally related to the condition and further research is needed on this issue. In studies Alexiou et al. [35] studied trace elements (zinc, cobalt, selenium, rubidium, bromine, gold) in the human placenta and in the liver of the newborn at birth, the authors note that the average concentration of essential trace elements (zinc, cobalt and selenium) were significantly higher in the liver than in the placenta, while interchangeable trace elements (rubidium, bromine, gold) were found in significantly higher concentrations in the placenta than in liver tissue.

Since the decrease in serum selenium level during pregnancy occurs monotonously, it is possible to calculate the optimal concentration of selenium in the serum of pregnant women: by the end of the first trimester—104 to 109 μg/l, the second—98 to 103 μg/l, to the third trimester—95 to 100 μg/l [39], these indicators can be used to assess the normal selenium status during pregnancy.

There is evidence of a connection with selenium deficiency weak labor activity, a significantly greater number of complications during delivery and lower development indicators of newborn [40]. The risk group for selenium deficiency is also pregnant women living in ecologically unfavorable regions and women with cardiovascular diseases [30]. In several case studies, the authors attributed the poor pregnancy outcome and reproductive failure, including recurrent pregnancy loss, with abnormal concentrations of selenium [12]. Other authors [13, 41, 42] also found a positive correlation between the increased risk of miscarriage in women with low selenium concentrations. Selenium reduction is a predictor of preterm birth and low birth weight, and correction of selenium deficiency during pregnancy eliminates the risk of low birth weight [28].

The concentration of selenium in the placenta at different periods of gestation was studied in children—residents of the Khabarovsk territory [43]. Estimating the central values of the studied microelements, the norm was understood as variants within one standard deviation; the boundaries of the norm were the values between the 25th and 75th centiles or ($M + 2\delta$), and the pathological values beyond these limits were considered. To calculate the extreme limits of normal values, the

Group	n	Average	m	Deviation interval
28–33 gw	14	0.217*,**,***	0.01	0.160–0.278
34–36 gw	17	0.341*	0.03	0.176–0.588
Fetal growth Retardation	14	0.271**	0.01	0.196–0.343
Control group	20	0.370***	0.04	0.233–0.626

Significant difference between 28 and 33 weeks and 34 and 36 weeks ($p = 0.0017$).
**Significant difference between 28 and 33 weeks and DLD don ($p = 0.0013$).*
***Significant difference between 28 and 33 weeks and full term ($p = 0.00003$).*

Table 2.
Selenium concentration (mg/kg) in the placenta at different gestational ages.

method of standard deviations was used—the lower limit—5 SD corresponds to 5 percentiles; the upper limit +5 SD—95 percentiles. Based on this, for all cases with values below 5 percentile, we consider a sign of a significant decrease in the parameter, and with values more than 95 percentile—it is stated as an increase (**Table 2**).

It is interesting to note that with the increase in gestation period, there is a more significant variability of selenium content, which does not occur in the group of children with intrauterine growth retardation syndrome.

Our findings suggest that passive transport through the placenta of selenium has insufficiently mature mechanisms of its regulation in the early stages of embryogenesis.

The study revealed that the level of selenium in the placenta has a dynamics similar to that in the hair of the mother and the newborn and in breast milk [29, 44].

The placenta is an organ that reflects the features of the course of the intrauterine period, the environmental situation, and the infectious background, and it is also an indicator of the content of vital for the normal carrying of the newborn trace elements.

4. Breast milk is the only natural source of selenium for infants

The only natural vitamin and mineral complex is breast milk; it contains all the necessary bioelements and is most adapted to the assimilation of the child. However, the composition of breast milk, both qualitatively and quantitatively, depends on the time of onset of labor, significantly different from normal indicators for premature termination of pregnancy and fetal growth retardation syndrome. Of particular importance for solving the problem of nutrition of small children [45] is a detailed explanation of the properties of breast milk and determining the value of each of its components [46].

Given that the child in the first months of life receives selenium exclusively from mother's milk and baby food does not always contain it in sufficient quantities, of particular importance is the provision of this micronutrient women, both during pregnancy and during lactation. The estimated need for selenium in premature infants is 20–25 μg/l in breast milk or infant formula (15 μg/l for full-term infants) [47, 48]. The main source of selenium for a child is breast milk, but many children, especially those receiving long-term inpatient treatment, are artificially fed and have lower values of selenium content [49]. In addition, the presence of selenium deficiency in the mother is a common cause of element deficiency in the newborn [50, 51]. The content of selenium in breast milk varies widely [9, 50, 52]. Thus, in the study of selenium in breast milk of women living in the United States, it was determined that the average selenium content is 18 μg/l, and the maximum level reached 60 μg/l [52]; a direct correlation was found between the level of selenium in milk and serum. In premature births, the milk of women living in New Zealand contains an average of 20 μg/l selenium [53]. The question of the needs of newborns in selenium is not finally resolved, but most researchers recommend enriching the mixture for children with selenium in an amount corresponding to its content in breast milk. But the supply of children with selenium in an amount equal to its content in breast milk is not equivalent, because in breast milk, mixtures contain different chemical forms of selenium with different levels of bioavailability and toxicity. Currently, baby foods mainly use selenite, which easily interacts with the ascorbic acid contained therein, forming an inactive elemental selenium; as a result, such products are inert against selenium [54]. Therefore, it seems natural to normalize the level of selenium in the mother feeding a newborn baby breast milk.

The content of selenium in breast milk during normal pregnancy and physiological childbirth by a conditionally healthy fetus and at low birth weight (**Table 3**) was determined in Khabarovsk [55].

When compared with the data of the WHO/MAGATE collaborative study [56], there were significant differences in the content of selenium in the milk of healthy women living in Khabarovsk (**Table 3**). The most pronounced deficiency of selenium in breast milk is preterm labor and fetal growth retardation syndrome; its level is 2 and 20 times lower than optimal, respectively, and from the first days of life of a small child, its insufficiency is formed with a significant ($p < 0.05$) maximum decrease in premature infants with a gestation period of 34–36 weeks and children with fetal growth retardation syndrome.

Selenium supply in newborns in the control group is optimal (**Table 4**).

At low weight, the actual provision of newborn children with selenium did not meet the standards of physiological needs of the body. The extremely low figures for the actual supply of selenium in natural feeding are detected in underweight children with the syndrome of fetal growth retardation and prematurity. Low level of selenium from the first days of life of a small child forms its negative balance. Given that the need for selenium at low weight is higher, and breast milk [55], as the only source

Trace elements	Control group, n = 20	28–33 weeks of gestation, n = 20	34–36 weeks of gestation, n = 20	Fetal growth retardation, n = 19	Literature data
Se	0.02 ± 0.007	0.01 ± 0.004**	0.009 ± 0.002*	0.001 ± 0.0002**	0.019 ± 0.001

**Significance of differences between the control group and childbirth at 34–36 weeks, $p < 0.05$.*
***Significance of differences between the control group, 28–33 weeks of labor, 34–36 weeks of labor, and fetal growth retardation syndrome, $p < 0.05$.*

Table 3.
Content of Se (mg/kg) in milk (M + m).

Groups	n	Se
The norm of physiological needs*		0.01
Childbirth 28–33 w	n = 20	0.003
Childbirth 34–36 w	n = 20	0.003
Childbirth a delay syndrome fetal growth retardation	n = 19	0.0003
Control group	n = 20	0.013

**Norms of physiological needs for energy and nutrients for different groups of the population of the Russian Federation were approved on December 18, 2008 (Mr 2.3.1.2432–08).*

Table 4.
Actual consumption of bioelements (mg/day).

Indicators	All subjects 13 months after delivery, n = 20	Lactation, 13 months after birth, n = 7	No lactation, 13 months after delivery, n = 13	Immediately after birth
Se	359.9 ± 18.9*	314.6 ± 20.2*	388.7 ± 25.1	524.6 ± 55.0*

**Significant differences (<0.05) between groups.*

Table 5.
Selenium content (μg/kg) in women's hair after childbirth (M + m).

of selenium for the newborn, contains less than optimal selenium, such a child is provided with the necessary trace element only 25–30% in premature birth and only 3% of the need for the birth of a child with delayed fetal development syndrome.

In the study of selenium content (**Table 5**) in the hair of 20 healthy women who gave birth to healthy children, a few months after birth, and an average of 13.05 + 0.54 months after birth (variability from 10 to 18 months) there were significant differences in the content of selenium in women's hair 13 months after birth depending on the duration of lactation. At the time of the study, seven women continued to feed their children with breast milk, the rest of the lactation ended 6 months or more ago [3].

In women with preserved lactation function by 13 months after birth, the level of selenium is significantly lower than in women who breastfed children for a shorter time.

Regardless of the presence of lactation, all subjects had significantly lower levels of selenium than immediately after birth. Thus, in the above observations, there is no recovery of selenium concentration after childbirth, typical for the content of selenium in blood serum. There were no significant differences in the content of selenium in the hair of women a year after childbirth with or without the use of vitamin complexes during lactation (368.9 + 22.13 and 336.4 + 38.7 μg/kg, respectively).

5. Children

Premature infants with low selenium levels have a higher rate of early neonatal morbidity [57]. The risk group for selenium deficiency includes children receiving long-term hemodialysis [18], with respiratory distress syndrome [58]; and children born in a state of chronic intrauterine hypoxia [59], with bronchopulmonary dysplasia [60]. In premature infants, selenium deficiency is associated with hypoxia and respiratory diseases [61]. In all these cases, the appointment of selenium is accompanied by a positive therapeutic effect. Confirmation of the diagnosis of chronic intrauterine hypoxia in the fetus can be a decrease in the content of selenium in the blood less than 25 μg/l [62]. Based on a multicenter randomized trial conducted by Darlow et al. in 2003, including a meta-analysis of other studies in this area, the additional use of selenium in the diet of preterm infants contributed to a reduction in the incidence of septic complications, which allowed the authors to recommend the use of selenium in the diet of preterm infants [53].

However, the optimal selenium supply of a newborn child is not currently determined; in the literature, there are different indicators of the norm from 65–75 μg/l [63] to 191 μg/l [64] in whole umbilical cord blood. In a study by Anya et al. [65] in the reference group of healthy newborns normal level of selenium was established in children younger than 1 month of life within the median 64 μg/L. Parfenova and Reshetnik [66] in the whole blood of preterm infants 27–32 weeks of gestation the level of selenium equal to 112.4 + 5.3 μg/l was determined.

Group	28–33 gw	34–36 gw	Fetal growth retardation	Control
n	23	34	21	20
Mother	0.3 ± 0.04* 0.02–0.7	0.4 ± 0.06* 0.02–1.4	0.4 ± 0.07* 0.02–0.8	0.6 ± 0.05 0.4–1.1
Children	0.5 ± 0.07* 0.04–0.9	0.6 ± 0.05* 0.08–1.01	0.2 ± 0.04*,** 0.01–0.6	0.8 ± 0.05 0.2–0.96

**Significant difference ($p < 0.05$) between study and control groups.*
***Significant difference ($p < 0.05$) when comparing the indicators of the fetal growth retardation group with the groups of 28–33 weeks and 34–36 weeks.*

Table 6.
Selenium content (mg/kg) in the hair of mother-newborn pairs in prematurity and delay syndrome fetal growth retardation (M ± m, min-max).

The concentration of selenium in serum is reduced in children with intraventricular hemorrhage [67], diseases of the respiratory system [68], and diseases of the gastrointestinal tract [69]. However, first of all, premature children are a risk group for selenium deficiency, aggravated by living in environmentally unfavorable conditions [3, 61, 70]. Many of them, especially those receiving long-term inpatient treatment, are artificially fed, and baby foods contain predominantly sodium selenite—an inorganic compound of selenium with high toxicity, low bioavailability and not always in sufficient quantities. Selenium obtained from mother's milk is better absorbed than selenium nutrient mixtures [71]. It is recommended to add selenium to the nutrition of mothers, as well as cows whose milk is used for the preparation of nutrient mixtures [63].

The mass of the newborn is largely determined by serum selenium and is inversely correlated. This also confirms the assumption that selenium is actively transported to the fetus in the quantities required by the embryo (**Table 6**).

Selenium levels in the hair of low-weight newborns and their mothers are statistically significantly different from the levels of selenium in the hair in the control group, with the highest degree of confidence established in the group with fetal growth retardation.

The deeper the immaturity and the lower the gestation period, the lower the level of selenium in the hair of premature infants and their mothers; the level of selenium in the hair of the group with fetal growth retardation decreased more than four times and was significantly lower than in the group of premature infants with the deepest immaturity.

Reduction of selenium in the hair of newborns increases the chance of oppression syndrome, muscle dystonia, and neutrophilia. RDS is often formed when selenium deficiency in the pair "mother-newborn" in prematurity.

Direct dependence of the average strength between the level of selenium in the hair and urine of pregnant women indicates unidirectional changes in the concentration of selenium.

Also in the process of gestation, there is a change in the content of selenium in serum and hair and a highly reliable feedback of average strength ($r = 0.538$) between the content of selenium in serum and hair of pregnant women.

6. Conclusion

Selenium deficiency in pregnant women can lead to an increase in the frequency and severity of early and late gestosis, fetal hypotrophy, and hypoxia, and has an impact on the duration of pregnancy and the rate of growth of fetal body weight. Selenium deficiency in pregnant women can lead to an increase in the frequency and severity of early and late gestosis, fetal hypotrophy and hypoxia, and also affects the duration of pregnancy and the rate of growth of fetal body weight, the formation of a deficiency in the fetus and newborn [15].

The study found a deficiency of selenium in women in preterm labor and is due not only to the lack of an element in the diet and the environment, but also an excess of selenium antagonists—Mn, Cd, Pb, and Fe.

The provision of selenium depends on the degree of maturity and physical development of newborns, decreases with fetal development syndrome, prematurity, artificial feeding, and hypoxia. The content of selenium in breast milk of prematurely born women provides only 25–30% of the needs of the element in premature infants. Breast milk of prematurely born women contains selenium three times less than in normal childbirth, which contributes to a high risk of developing alimentary-dependent conditions in premature infants [72].

Author details

O.A. Senkevich[1]* and Y.G. Koval'skiy[2]

1 Department of Additional Professional Education, The Institute of Continuing Professional Education and Accreditation of Far Eastern Medical University Ministry of Health of Russia, Khabarovsk, Russia

2 Department of Biological Chemistry, Clinical Laboratory of Far Eastern Medical University the Ministry of Health of Russia, Khabarovsk, Russia

*Address all correspondence to: senkevicholga@yandex.ru

References

[1] Tutelian VA. Nutrition: Contemporary issues. Health Care of the Russian Federation. 2008;**1**:8-9

[2] Golubkina NA, Alfthan GV. The human selenium status in 27 regions of Russia. Journal of Trace Elements in Medicine and Biology. 1999;**13**:15-20 (in Russian)

[3] Senkevich OA, Golubkina NA, Kowalski YG, Sirotina ZV, Iskrenok GV. Security selenium of residents of the Khabarovsk territory. Far East Medical Journal. 2009;**1**:82-84 (in Russian)

[4] Thomson CD. Assessment of requirements for selenium and adequacy of selenium status: A review. European Journal of Clinical Nutrition. 2004;**58**:391-402

[5] Kodentsova VM, Vrzhesinskaya OA, Reznik DV, Nikityuk DB, et al. The Security of Russia's population with micronutrients and possibilities of its correction. Status of the problem. Voprosy Pitaniia. 2017;**86**(4):113-124. DOI: 10.24411/0042-8833-2017-00067. (in Russian)

[6] Al-Saleha I, Billedoa G, El-Dousha I, et al. Selenium and vitamins status in Saudi children. Clinica Chimica Acta. 2006;**368**(1):99-109

[7] Schomburg L. Dietary selenium and human health. Nutrients. 2016;**9**:1. DOI: 10.3390/nu9010022.17

[8] Nefedova ND, Antonov AR, Efremov AV. Selenium and reproductive function in women of Novosibirsk. Siberian standard of living: Ecology of nutrition. In: Abstracts. Scientific and Practical Conference. Novosibirsk; December 2-3, 1998. p. 212 (in Russian)

[9] Shagova MV. Hygienic assessment of selenium supply in pregnant women and children of Russia [dissertation]. PhD Biol. Sciences; Moscow, 2000. 18p (in Russian)

[10] Mistry HD, Kurlak LO, Young SD, Briley AL, Broughton Pipkin F, Baker PN, et al. Maternal selenium, copper and zinc concentrations in pregnancy associated with small- for-gestational-age infants. Maternal & Child Nutrition. 2012:48

[11] Nebyshinez LM, Silyava V, Kamyshnikov VS. Eatures of trace element composition of blood in pregnant women at early gestation. Medicine (Minsk). 2004;**4**:36

[12] Ajayi GO. Selenium concentration in first trimester abortion in Nigerian women. Trace Elements and Electrolytes. 2004;**21**:1-3

[13] Zachara BA, Dobrzynski W, Trafikowska U, Szymanski W. Blood selenium and glutathione peroxidases in muscarriagr. BJOG : An International Journal of Obstetrics and Gynaecology. 2001;**108**:244-247 57

[14] Kieliszek M, Blazejak S. Current knowledge on the importance of selenium in food for living organisms: A review. Molecules. 2016;**21**:5. DOI: 10.3390/molecules21050609

[15] Kantola M, Purkunen R, Kröger P, Tooming A, Juravskaja J, Pasanen M et al. Selenium in pregnancy: Is selenium an active defective ion against environmental chemical stress. Environmental Research. 2004;**96**(1):51-61

[16] Golubkina NA, Papazyan TT. Selenium in Nutrition. Plants, Animals, Man. White City: Publishing; 2006. 250p. ISBN: 5-98467-002-X (in Russian)

[17] Reshetnik LA, Parfenova EO. Biogeochemical and clinical significance of selenium for human health. Trace

Elements in Medicine. 2001;**2**:2-8 (in Russian)

[18] Allman MA, Stewart PM, Tiller DJ. Energy supplementation and the nutritional status of hemodialysis patients. The American Journal of Clinical Nutrition. 1990;**51**:558-562 76

[19] Kostyrko DG, Senkevich OA. Availability of selenium depending on the mode and quality of nutrition students of Khabarovsk city. Actual problems of neonatology and Pediatrics: Collection of scientific works. In: Proceedings of the Far Eastern Scientific and Practical Conference with International Participation; Khabarovsk; 2017. pp. 118-121 (in Russian)

[20] Klapes T, Cavar S, Kasac Z, et al. Selenium in placenta birth weight in normal but not intrauterina growth restriction pregnancy. Journal of Trace Elements in Medicine and Biology. 2008;**22**(1):54-58

[21] Capellia R, Mingantia V, Seminob G, Bertarini W. Presence of mercury (total and organic) and selenium in human placentae. The Science of the Total Environment. 1986;**48**(1-2):69-79

[22] Lorenzo Alonso MJ, Bermejo Barrera A, Cocho de Juan JA, Fraga Bermúdez JM, Bermejo Barrera P. Selenium levels in related biological samples: Human placenta, maternal and umbilical cord blood, hair and nails. Journal of Trace Elements in Medicine and Biology. 2005;**19**(1):49-54

[23] Lorenzo Alonso MJ, Bermejo Barrera A, Cocho de Juan JA, et al. Selenium levels in biological samples: Human placenta, maternal and umbilical cord blood, hair and nails. Journal of Trace Elements in Medicine and Biology. 2005;**19**(1):49-54

[24] Tutelian VA, Spirichev VB, Shatnyuk LN. Correction micronutrient deficiency—The most important aspect of the concept of healthy eating in Russia. Voprosy Pitaniia. 2014;**83**(1):3 (in Russian)

[25] Ward NI, Watson R, Bryce-Smith D. Placental element levels in relation to fetal development for obstetrically "normal" births: A study of 37 elements. Evidence for effects of cadmium, lead and zinc on fetal growth, and for smoking as a source of cadmium. The International Journal for Biosocial Research. 1987;**9**:63-81

[26] Artemyeva EK, Setko NP, Saprykin VB, Vekker IR. Concentration of trace elements in the mother-placenta-fetus system in areas with different levels of anthropogenic load. Trace Elements in Medicine. 2004;**5**:1-3 (in Russian)

[27] Zadrozna M, Gawlik M, Nowak B. Antioxidants activities and concentration of selenium, zinc and copper in preterm and developmental delay human placentas. Journal of Trace Elements in Medicine and Biology. 2009;**23**:144-148

[28] Bogden JD, Kemp FW, Chen X, et al. Low-normal serum selenium early in human pregnancy predicts low birth weight. Nutrition Research. 2006;**26**(10):497-502 59

[29] Senkevich OA. Microelement imbalance in the formation of pathology of small-weight newborns in the far East [Text autoref. dis]. Khabarovsk; 2009. 42p

[30] Golubkina NA, Shagova MV. Assessment of the level of consumption of selenium by pregnant women in some regions of Russia and Ukraine. Nutrition of children: The Twenty-First Century. In: Proceedings of All-Russian Congress with International Participation; 2000. p. 28 (in Russian)

[31] Iyengar GV, Raphuman A. Placenta as a 'dual' biomarker for monitoring

fetal and material environment with special reference to potentially toxic trace elements: Part 2: Essential minor, trace and other (non-essential) elements in human placenta. The Science of the Total Environment. 2001;**280**(1-3):207-219

[32] Kantola M, Kroger P, Purkunen R, Tooming A, Juravskaja J, Pasanen M, et al. Accumulation of cadmium, zinc and copper in maternal blood and developmental placental tissue differences between Finland, Estonia and St. Petersburg. Environmental Research. 2000;**A83**:54-66

[33] Bratter P, Nefreti de Bratter VE, Recknagel S, Brunetto R. Maternal selenium status influences the concentration and binding pattern of zinc in human milk. Journal of Trace Elements in Medicine and Biology. 1997;**11**:203-209 83

[34] Kantola M, Vartiainen T. Changes in selenium, zinc, copper and cadmium in human milk during the time when selenium has been supplemented to fertilizers in Finland. Journal of Trace Elements in Medicine and Biology. 2001;**15**:11-17 90

[35] Alexiou D, Grimanis AP, Grimani M, Papaevangelou G, Koumantakis E, Papadatos C. Trace elements (zinc, cobalt, selenium, rubidium, bromine, gold) in human placenta and newborn liver at birth. Pediatric Research. 1977;**11**(5):646-648

[36] Kulakov VI, Antonov AG, Baybarina EN. Problems and prospects of nursing children with extremely low body weight at the present stage. The importance of selenium in children's health and reproduction. Russian Bulletin of Perinatology and Pediatrics. 2008;**4**:8. DOI: 10.5772/intechopen.88864 (in Russian)

[37] Magos L, Webb M. The interactions of selenium with cadmium and mercury. Critical Reviews in Toxicology. 1980;**571**:1-42

[38] Flora SJ, Singh S, Tandon SK. Role of selenium in protection against lead intoxication. Acta Pharmacologica et Toxicologica. 1983;**53**:28-32

[39] Golubkina NA, Alfthan G. Selenium status of pregnant women and newborns in The Former Soviet Union. Biological Trace Element Research. 2002;**89**(1):13-23

[40] Gmoshinskaya MV, Gmoshinsky IV, Shilina NM, et al. Alimentary correction of selenium deficiency in pregnant women. Ways to improve the effectiveness of medical care for children. In: 10th Congress of Pediatricians of Russia; 2005. p. 119 (in Russian)

[41] Kocak I, Aksoy E, Ustun C. Recurrent spontaneous abortion and selenium deficiency. International Journal of Gynecology & Obstetrics. 1999;**65**:79-80

[42] Nourmohammadi I, Mehdizadeh A, Mandegar M. The status of zinc and selenium in Iranian women during normal pregnancy and in case of recurrent miscarriage. Trace Elements in Medicine. 2005;**4**(6):23-26

[43] Kowalski YG, Senkevich OA, Golubkina NA, Sirotina ZV. Security selenium of residents of the Khabarovsk territory. Far East Medical Journal. 2009;**1**:82-84 (in Russian)

[44] Komarova ZA, Senkevich OA, Sirotina ZV. Essential microelements of female breast milk in preterm labor. In: Proceedings of the 4th Congress of Perinatal Medicine Specialists; 2009. p. 31 (in Russian)

[45] Senkevich O, Plotonenko ZA, Lemeshchenko OV, et al. Adherence of mothers to recommendations on proper nutrition of children and "non-baby"

products in the diet of young children. Questions of Children's Dietetics. 2018;**16**(4):20-28 (in Russian)

[46] The National Program for the Optimization of Nutrition of Children Under the 1 Year in the Russian Federation et al. The Union of Pediatricians of Russia. 2nd ed. 2019. 36p (in Russian)

[47] Bath N, Cobb K, Dwyer J, et al. Recommendations on nutrition of infants and young children (Program Start Healthy—Grow healthy from the first days). Questions of Modern Pediatrics. 2007;**5**(1):115-129 (in Russian)

[48] Senkevich OA, Komarova ZA. Selenium deficiency in preconception and lactation: Consequences and ways of correction. Far East Medical Journal. 2016;**2**:42-46 (in Russian)

[49] Litov RE, Combs GF Jr. Selenium in pediatric nutrition. Pediatrics. 1991;**87**(3):339-351

[50] Hannan MA, Dogadkin NN, Ashur IA, Markus WM. Copper, selenium and zinc concentration in human milk during the first three weeks of lactation. Journal of Trace Elements in Medicine and Biology. 2007;**21**:49-52

[51] Kim SY, Park JH, Kim EA, Lee-Kim YC. Longitudinal study on trace mineral compositions (selenium, zinc, copper, manganese) in Korean human preterm milk. Journal of Pediatric Surgery. 2012;**47**(4):760-771

[52] Shearer TR, Hadjimarcos DM. Geographic distribution of selenium in human milk. Archives of Environmental Health. 1975;**30**:230-233

[53] Sluis KB, Darlow BA, George PM, et al. Selenium and glutathione peroxidase level in premature infant in a low selenium community (Christchurch, New Zealand). Pediatric Research. 1992;**32**(2):189-194

[54] Shrimpton DH. Micronutrients and their interaction. Russian Open Medical Journal. 2008;**7**:453-457 (in Russian)

[55] Komarova ZA, Senkevich OA, Sirotina VZ. The balance of essential trace elements (zinc, copper, selenium) in breast milk—The basis of prevention of pathological conditions of the newborn. In: The Collection of Materials of the 1st Congress of Pediatricians of the Far East "Actual Problems of Protection of Motherhood and Childhood on the Modern Stage." Khabarovsk; 2010. pp. 133-135 (in Russian)

[56] WHO/MAGATE. Trace Elements in Breast Milk: Report of the Joint Collaborative Study. 1991. 135p

[57] Burjonrappa SC, Miller M. Role of trace elements in parenteral nutrition support of the surgical neonate. Journal of Pediatric Surgery. 2012;**47**:760-771

[58] Singleton TA, Clemson LA, Gore DC. Supportive care in acute respiratory distress syndrome. Seminars in Thoracic and Cardiovascular Surgery. 2006;**18**(1):35-41

[59] Klyuchnikov SO, Demekhina MF, Demin VF. Mineral composition of blood serum of newborns in the early neonatal period. Pediatrics. 1995;**3**:28-32 (in Russian)

[60] Jobe AH, Bacalari E. Broncho-pulmonary displasia. American Journal of Respiratory and Critical Care Medicine. 2001;**163**:7

[61] Netrebenko OK. Copper and selenium in nutrition of premature infants. Russian Pediatric Journal. 2005;**1**:25-27 (in Russian)

[62] Klyuchnikov SO, Deshchekina MF, Demin VF. The content of macro- and

microelements in the blood serum of newborns as one of the criteria for assessing the state at birth. Pediatrics. 1994;**6**:53-55 (in Russian)

[63] Tutelyan A, Maso VK, Shiro LI. The value of selenium in human nutrition. Gynecology. 2002;**2**:12-14 (in Russian)

[64] Iijima K, Otake T, Yoshinaga J, Ikegami M, Suzuki E, Naruse H, et al. Cadmium, lead, and selenium in cord blood and thyroid hormone status of newborns. Biological Trace Element Research. 2007;**119**(1):10-18 54

[65] Muntau AC, Streiter M, Kappler M, Roschinger W, Schmid I, Rehnert A, et al. Age-related reference values for serum selenium concentrations in infants and children. Clinical Chemistry. 2002;**48**:555

[66] Parfenova EO, Reshetnik LA. On the correction of selenium status of premature infants. In: Proceedings of the 1st International Practical Scientific-Practical Conference. Orenburg; 2004. pp. 127-129 (in Russian)

[67] Paul DA, Leef KH, Stefano JL, Bartoshesky L. Low serum thyroxine on initial newborn screening is associated with intraventricular hemorrhage and death in very low birth weight infants. Pediatrics. 1998;**101**:903-907

[68] Lands LC. Nutrition in pediatric lung disease. Pediatric Respiratory Reviews. 2007;**8**(4):305-312

[69] Thomas AG, Miller V, Shenkin A, et al. Selenium and glutathione peroxidase status in paediatric health and gastrointestinal disease. Journal of Pediatric Gastroenterology and Nutrition. 1994;**19**(2):213-219

[70] Klenova NA. Metabolism of the main nutrients in different types of feeding premature infants 1 month of life [dissertation]. Moscow; 2001. 21p (in Russian)

[71] Kon I, Kopytko MV. The value of selenium in maintaining the health of children. Children's Doctor. 2000;**5**: 38-40 (in Russian)

[72] Sirotina ZV, Senkevich OA. Microelement imbalance in the formation of pathology of low-weight newborns in the Far East. Far East Medical Journal. 2010;**4**:47-50

Chapter 3

Uptake, Metabolism and Toxicity of Selenium in Tropical Plants

Abiodun Humphrey Adebayo, Omolara Faith Yakubu and Osarobo Bakare-Akpata

Abstract

Selenium is a mineral element that is essential for both animal and humans and can also serve as an environmental toxicant. A narrow margin exists between an ideal and toxic intake of selenium. It is a useful microelement existing in minute amounts in animals, plants, microorganisms and humans. Although it is beneficial to both animals and humans as an antioxidant, it can be toxic at high concentrations as a result of it competing and replacing sulfur in amino acids leading to inappropriate folding of protein and eventually creating a nonfunctional protein and enzymes. Selenium exists in organic forms as SeMet and SeCys and inorganic forms as selenide, selenite and selenite in the environment. It is translocated in plants via the sulfate transporters in the plasma membrane of the plant root. Its translocation and distribution however depends on the plant species, their different developmental phases, forms, concentration and other physiological conditions like pH. Inorganic selenium is first converted to selenite via the action of two different enzymes (ATP sulfurylase and APS reductase), selenite is further converted to selenide by sulfite reductase. Selenide eventually couples with O-acetyl serine via the action of cysteine synthase to form SeCys. SeCys can either be methylated to methyl-SeCys through the action of selenocysteine methyltransferase or to elemental selenium via SeCys lyase or converted by a series of enzymes to selenomethionine. Selenium toxicity or Selenosis can occur when the optimal concentration of selenium is exceeded. Two major mechanism of selenium toxicity exists; either by induction of oxidative stress or malformation of selenoproteins. Selenium uptake, metabolism and toxicity in tropical plants are hereby discussed in this chapter.

Keywords: selenium, distribution, toxicity, tropical plants

1. Introduction

Selenium (Se) is a widely distributed trace metalloid found in the crust of the Earth. Jacob Berzelius, Chemist first isolated selenium in 1817 and it has been known for its toxic effect. However, in 1957, some importance of selenium was discovered. It is mostly linked to sulfur and an essential nutrient for human, animals and microorganisms. Many enzymes such as thioredoxin reductase and glutathione peroxidase are mostly composed of selenium which helps the enzymes to perform roles like reproduction, tumor prevention and antioxidation [1]. Selenium can also promote growth of lettuce seedlings by delaying senescence [2]. It can be toxic at large concentrations and can lead to pro-oxidative reactions. Also deficiency of selenium can occur in soils where selenium bioavailability is low leading to health risks for animals and humans. Supplementing fertilizers with sodium selenate has been shown to improve the food chain from soil to animals and then to humans [3]. The recommended daily intake of selenium should be adhered to for maximum utilization of its benefits. Within the plant and soil environment, selenium is converted to another chemical form [4]. The metabolism and mechanisms through which plants cope with high selenium concentrations are explained by the transformations of selenium form one form to another. The bioavailability, biotransformation, speciation, metabolism and functions all have great implications for both human and animal health. This chapter presents the physicochemical properties of selenium, its sources in the environment and locations, role of selenium in the body, metabolism, uptake and accumulation in plants.

2. Physicochemical properties of selenium

Selenium being a metalloid from the same family of sulfur and oxygen, has it name derived from the word "Selene" that is, moon goddess since it is mostly linked to tellurium [5]. It has six isotope coexisting in nature with mass numbers 74, 76, 77, 78, 80 and 82 [6]. It is similar to sulfur in terms of bond energies, oxidation state, atomic size and ionization potentials [7]. Selenium possesses properties of both non-metal and metal hence it is referred to as a semi metal. It is considered stable as it does not oxidize at room temperature. It produces selenium dioxide and blue flame when it burns which is followed by an unpleasant smell. Selenium can form compounds with elements (fluorine, bromine, hydrogen and phosphorus) having a close analogy to those of sulfur [8, 9]. It has a lower affinity for oxygen than sulfur with only two oxides known; SeO_3 and SeO_2. Combustion of selenium in air produces dioxide which dissolves in water to give selenious acid (H_2SeO_3); a solution that can oxidize most metals except platinum, palladium and gold [10–20].

Selenic acid (H_2SeO_4) is a hygroscopic diacid with a higher oxidizing potential than H_2SO_4. It is produced by the reaction of oxidizing agents such as chlorine, fluorine, bromine with Se, SeO_2, H_2SeO_3 in the presence of H_2O. Reaction of selenium with hydrogen and reaction of metal selenides with acids (or water) releases hydrogen selenide (H_2Se), a highly reactive compound. At about 160°C, it starts to decompose to Se and H_2, it also forms a deposit of red selenium in moist air [21].

3. The physical and chemical forms of selenium

Selenium exists in nature and in organisms in organic and/or inorganic forms. The organic form includes selenocysteine (Secys) and selenomethionine (Semet), while the inorganic forms includes selenate (SeO_4^{-2}), selenide (Se^{-2}), selenite (SeO_3^{-2}) and selenium (Se) (**Figure 1**) [22].

Selenium exists in a solid state at room temperature and can take up various physical forms [8, 23]. Precipitation from aqueous solution produces amorphous selenium (red brick powder) with a density of 4.26 with photoconductive properties. At very high temperature between 110 and 180°C, the color turns gray, this is a variety of selenium that is thermodynamically stable and it is obtained by cooling liquid selenium hence it is used for its semiconducting properties.

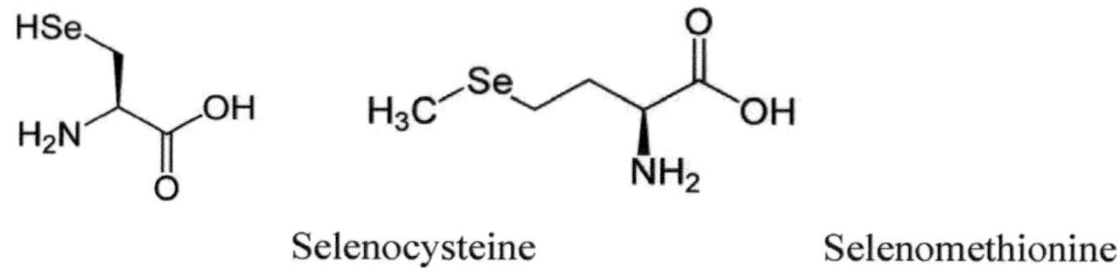

Figure 1.
Selenocysteine and selenomethionine.

4. Sources of selenium in the environment and its location

4.1 In soils

Selenium occurs in soils via the erosion of rocks that contains selenides and selenites associated with sulfide minerals with mass fractions less than 1 mg/kg. Selenium is mostly found in soils either in its organic form or elemental selenium like selenite salts and ferric selenite. The common forms of selenium in soils are the anionic forms like Selenite (SeO_3^{2-}) and selenate (SeO_4^{2-}) and they are soluble and potentially toxic. Organic forms of selenium in soils are mostly from plants decomposition [24, 25].

The selenium content in soil varies depending on the organic matter, soil texture along with the level of rainfall. The rate of assimilation of selenium by plant is further influenced by the physicochemical factors of the soil, such as microbial activity, pH and redox status. The concentration of selenium in the soil varies from 0.1 to 0.7 mg kg^{-1}. Tropical soils have a concentration between 2 and 4.5 mg kg^{-1}, while clay soils are between 0.8 and 2 mg kg^{-1}. Granites and volcanic soils are usually low in selenium while the soils around mountains are rich in selenium. Normally, selenium is more concentrated in soils of the driest regions of the world. The toxic effect of selenium on animals occurs on these soils [26, 27]. The rate of soil acidity determines the amount of selenium in crops and plants; more selenium is released in alkaline soil than in acidic ones. In alkaline soils, selenite undergoes oxidation into a more soluble form that is easily assimilated by plants (selenite). However, in acidic soils, selenite combines with iron hydroxide making it to be permanently fixed by the soil [28].

4.2 Plant sources

The concentration of selenium in plants is dependent on the level of selenium levels in the soils. The physiological conditions and species of the plant also determine how the selenium are taken in and distributed by the plant root. The aerial silks normally contain about 80% selenite and 65% selenate [29]. Forages contain selenium from about 0.2 to 0.6 ppm and livestock are at risk of selenium poisoning [30]. In arid regions of the United States and China, some plants contain very high selenium content, as high as 10,000 ppm [4]. Some species of *Astragalus* usually accumulates very high level of selenium making them toxic to animals [31]. Wheat plants normally store selenium in their seeds in form of selenomethionine with varying levels depending on the environment [27, 32]. Plants assimilate selenite more than selenite. Selenate and selenite share similar chemical features with sulfur hence they both undergo metabolism via the same route (in the choloroplast). The first reaction is when selenite is activated by ATP sulfurylase-adenosine 5′-phosphoselenate which is followed by its reduction to selenite by adenosine

5′-phosphosulfate reductase and finally the selenite is further reduced by the action of glutathione into selenide. Depending on the rate of selenium accumulation by plants, two mechanism of metabolism exist.

4.3 Selenium in water

Selenium in water is initiated from ambience deposits or drainage in the soil, its concentration varies but does not exceed 9 mg L^{-1}. The World Health Organization recommends that the content of selenium in water for consumption should not exceed 10 μg L^{-1} [24]. When farm lands are supplemented with fertilizers, selenium content increases. Selenate sodium and selenide are mostly found in surface waters, while freshwater contains majorly selenite.

4.4 Food and feed sources of selenium

The content of selenium in vegetables and grains is largely dependent on selenium found in the soil. Vegetables like beans, peas and carrots can contain up to 6 mg g^{-1} of selenium, while onions contain more. However fruits generally contain a very low level of selenium, but nuts with high protein levels are also known for their high selenium concentration [33–35].

5. Role of selenium in the body

Selenium, an important component of selenoprotein, plays diverse biological roles ranging from antioxidant defense to synthesis of DNA to reproduction. Various metabolites formed from selenium could also play a role in the prevention of carcinogenesis. It could also improve tolerance and recuperation thus slowing down the aging process [36, 37].

5.1 Selenoprotein

5.1.1 Glutathione peroxidase (GPx)

An antioxidant, whose primary role is to counteract the effect of hydrogen peroxide and other hydroperoxides in the body. GPx exists in about eight different forms grouped according to their features. They differ by mode of action and site of action. They work alongside vitamin E to protect cells from accumulated H_2O_2 hence they ensure the integrity of the cell wall. The first four forms of GPx enzymatic activity are directly proportional to the intake of selenium. Hence there is a correlation between oxidative stress and lack of selenium in the body [38–40].

Glutathione peroxidase-1 (GPx-1) occurs mostly in the liver, erythrocytes, lungs and kidneys. Deficiency in selenium affects the activity of GPx-1. Glutathione peroxidase-2 (GPx-2) protects against oxidation and it occurs mostly in the gastrointestinal tissues and the liver [41]. GPx-3 is found in the plasma, heart, kidneys, liver and it covers over 20% of the plasma selenium. It reduces the level of hydroperoxides [42]. GPx-4 is located in the mitochondria, nucleus and cytosol with its highest activity in the testes [43]. In addition to its antioxidant role, it prevents occurrence of peroxidation on the membrane. It is involved in the conversion of cholesterol and its ester into non-toxic derivatives and also prevents oxidation that can lead to DNA damage. The role of GPx-5 is still unknown but it is found in the embryo, while the other GPx: 6, 7, 8 are less studied [44].

5.2 Roles of selenium in the immune response

The lymph nodes and the liver contain high amount of selenium which helps to brace up the formation of antibodies and increase the functioning of the helper T cells and cytotoxic NK cells. It also stimulates the migration of the phagocytic cells [8, 45]. Metabolites of selenium such as GPx-1 and thioredoxin reductase have also been implicated in the inflammatory and immune responses although the mode of action is not fully known [46, 47]. Deficiency of selenium in the endothelial cells reduced the production of prostaglandins. In addition, it was reported that dairy cows deficient in selenium had low production of blood neutrophils hence their ability to kill a pathogen was nullified [47]. A rapid production and differentiation of CD4+ and T cells were observed in subjects who ingested selenium leading to increased poliovirus clearance [48].

5.3 Cancer and cardiovascular disease

Davis et al. [49] illustrated the correlation between selenium and cancer. Their studies showed that one of the factors that promote cancer is selenium deficiency. The authors discovered and reported that high selenium levels reduced the risk of cancer by 4–6 times when compared with low intake levels of selenium (<50 μg/mL). Populations with a status of very low selenium signified more protection against lung cancer [49]. Liver cancer was reduced by 30% in a community whose diet was enriched with selenite supplements [49]. In a Nutritional Prevention of Cancer trial, a daily intake of selenium (200 μg) for a period of 7 years lowered the occurrence of prostate cancer among the participants [50]. In a similar fashion, selenium anticancer potential was observed in rodents where the enzyme that converts selenomethione to methylselenol was 700 times higher in the rodents [49].

5.4 Role of selenium in reproduction

Studies have reported the association of selenium in animal and human reproduction. Selenium plays unique roles in fertility, placenta retention, synthesis of sperm and testosterone. Consumption of selenium deficient diet has been linked to poor growth and reduced fertility [51]. Changes in the luteinizing hormone receptors of Leydig cells observed selenium deficiency as it affects secretion of testosterone (Thomson and Robinson; [52]). Several studies have reported the protective effect of selenium in cadmium-induced toxicity. Selenium plays a role in inhibiting the growth of cancer cells in prostate cancer subjects by inhibiting RNA, DNA and protein synthesis [53]. Selenium has been reported to impact the entire morphology of the testis [54]. Selenium has been shown to increase fertility in dairy sheep [55]. Pastures with very low selenium levels were found to have increased fertility when administered selenium supplements [40]. Such an increase was not observed when the supplement was replaced with vitamin E. Administering selenium supplements to pregnant Ewes increased the rate of lamb's survival during the first 10 days [37]. Injection of selenium decreased the formation of ovarian cyst in cows with deficient diet [56]. Prolapse of the cervix was attributed to selenium deficiency, while red blood cells with low selenium concentrations were reported in women undergoing uncontrolled abortions [57]. Fertility and sperm was improved after consumption of selenium in a study conducted in Scotland [57].

6. Metabolism of selenium

6.1 Transformation, absorption and transport

Glutathione plays a major role in the metabolism of selenium; it is involved in reduction reactions where selenite is converted to hydrogen selenite (H_2Se) which further releases the selenium for selenoprotein synthesis. The hydrogen selenide undergoes several methylations to finally arrive at formation of trimethylselenonium ion [$(CH_3)_3Se^+$] [57]. The rate of absorption of selenite in sheep much lower (29%) when compared with pork (80%) while selenate and selenomethionine have greater absorption rate in poultry animals. This is a result of reduction of selenite that is not available in ruminants [40]. Absorption occurs mostly inside the caecum and duodenum by active transport via a sodium pump. The mode of action differs depending on the specific form of selenium. Adsorption could be by simple diffusion e.g. selenite or by cotransport while the selenomethionine are absorbed via the amino acid uptake method [52, 58]. Elements like lead, sulfur and arsenic slows down the rate of absorption of selenium either through competing with selenium or by formation of complexes that are not capable of being assimilated [59]. Selenium level in the hepatocytes determines the level of absorption in the intestine. Erythrocytes take up selenium rapidly and it undergoes reduction by glutathione reductase and finally

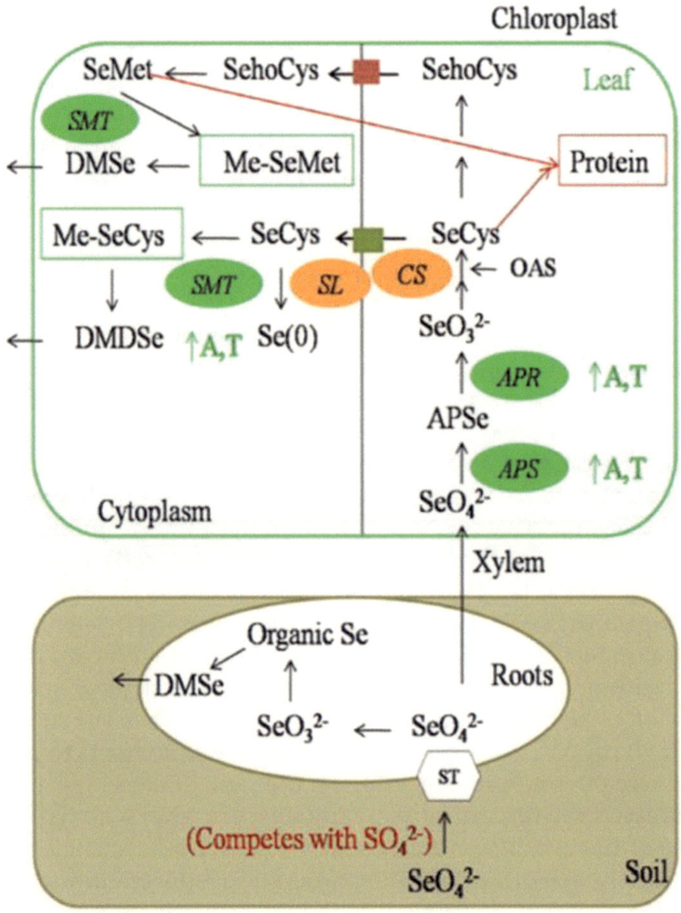

Figure 2.
Selenium metabolism in plants.

transported in the form of selenide in the plasma to the liver [60]. Selenium can also bind to either α and β globulins and can be transported in the form of selenoprotein by the blood [40].

6.2 Selenium metabolism in plants

Selenium is transported in plants via the sulfate transporters in the plasma membrane of the root due to its chemical similarity to sulfur [13, 14]. It is further transported to the leaves and undergoes metabolism through the sulfur assimilation pathway either to a selenium methionine (SeMet) or a selenium cysteine (SeCys).

Inorganic selenium is first converted to selenite via the action of ATP sulfurylase (APS) and APS reductase (APR). The hydrolysis of adenosine triphosphate to adenosine phosphoselenate which is then reduced to selenite is catalyzed by APS and APR, respectively [13]. Sulfite reductase then converts selenite to selenide although glutaredoxins or glutathione can also reduce this step in plants [17]. The selenide couples with O-acetyl serine (OAS) to form selenium cysteine (SeCys) by cysteine synthase. The selenium cysteine can either undergo methylation to methyl-SeCys by selenocysteine methyltransferase or converted to selenium or selenomethionine by SeCys lyase or other enzymes, respectively. Sulfur analog of selenium can then be methylated and undergo vaporization to a non-toxic form in the atmosphere (**Figure 2**) [15].

7. Selenium uptake and accumulation in plants

7.1 Selenium uptake

Selenium exists as both as organic (seleniumcysteine (SeCys) and selenium-methionine (SeMet)) and inorganic (selenate (SeO_4^{2-}), selenide (Se^{2-}), selenite (SeO_3^{2-}) and selenium (Se)) [61, 62]. The various species of plants, their developmental phases, type and concentration of selenium, the soil pH and its salinity determines the uptake and distribution of selenium in plants [14, 16]. Selenate is the most bioavailable form of selenium in agriculture and it is also more soluble in water than selenite [63]. Selenite is found in acidic soils, whereas selenate is found in alkaline soils [14]. According to Kikkert and Berkelaar [64], the rate of translocation of selenate is higher than that of selenium methionine while that of selenium methionine is greater than that of selenite or selenium cysteine by studying the translocation factors. Uptake of selenium is mostly carried out by transporters in the cell membrane of the root; selenate is transported by sulfate transporters while selenite is transported by phosphate transporters [14, 65]. The plants nutritional state determines the choice of the transporters [66]. Transporters for selenium decrease under extreme sulfate concentrations while the inducible transporters show greater affinity for sulfate than selenate than the constitutive ones [66]. Lack of sulfur and phosphorus in *Triticum aestivum* enhanced the uptake of selenium [14].

7.2 Se accumulation in plants

Selenium are usually concentrated in younger leaves during the period of seedling and tend to accumulate in the vacuoles of plant cells and are discharged via sulfate transporters in the tonoplast [67–69]. Different categories of selenium accumulation exist in plants: non-accumulators, secondary accumulators and hyperaccumulators (**Figure 3**; [61]). The non-accumulators are plants that

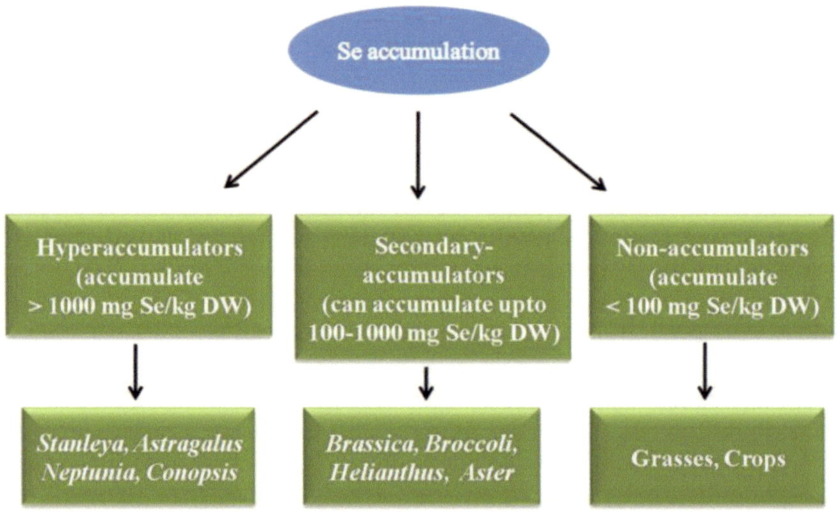

Figure 3.
Classification of plants based on selenium accumulation.

accumulate lesser than 100 mgSe/Kg of their DW, cannot survive on selenium-rich soils and volatilize selenium in form of dimethylselenide (DMSe), for example, grasses [61]. The secondary accumulators show no sign of toxicity at 1000 mgSe/Kg DW, for example, *Camelina, Brassica napus.* Finally, the hyperaccumulators by their name accumulates greater amount of selenium (>1000 mgSe/Kg DW), flourish in selenium-rich areas of the world and they release selenium as dimethyldiselenide, such plants includes *Xylohiza* and *Conopis.*

8. Beneficial effects of selenium in plants

8.1 Metabolic importance of selenium for plants

Selenium is a non-metal that is toxic at high concentration in plants but plays an important role at lower concentrations in them. Selenium increases plant growth when triggered by ultra-violet irradiation. The selenium and UV light share a synergistic relationship in the absence of selenium the UV light is capable of damaging the plant but in the presence of Se, plant growth is increased [70]. Selenium also functions as an antioxidant in plants. This antioxidant activity is also responsible for the improved growth in the plant. Selenium exerts its antioxidant activity by alleviating lipid peroxidation through GSH-Px activity. Se induced two selenoproteins, which are the thioredoxin reductase and GSH-Px these enzymes protect the plant from oxidative stress (Djanaguiraman et al., 2010). Se protects some plants from abiotic stress at low concentrations [71].

8.2 The functions of selenium in plants

Selenium functions differently in various plants;

i. It induces starch accumulation in chloroplast [72]

ii. Promotes germination [73]

iii. Increases respiratory potential [74]

iv. Improves nitrogen assimilation [75]

v. Increases the shoot and dry matter production (Djanaguiraman et al., 2010)

9. Selenium toxicity in plants

Selenium toxicity in plants arises from an increased concentration of selenium beyond the optimum threshold [70, 76]. High concentrations of Se in plant root can cause them to exhibit symptoms of injury, stunting of growth, chlorosis, withering and drying of leaves, decreased protein synthesis and premature death of the plant [77, 78]. Selenium toxicity is caused by two mechanisms: the first by inducing oxidative stress and second by malformed selenoproteins [70] plant from oxidative stress (Djanaguiraman et al., 2010). Se protects some plants from abiotic stress at low concentrations [71].

9.1 Toxicity due to malformed selenoproteins

Malformed selenoproteins result from the substitution of seCys/seMet into the protein chain in place of Cys/Met, these se amino acids are unstably unfavorable to protein functioning. Cysteine plays a primary role in the structure and function of a protein chain, disulfide bond formation, chemical catalysis and also functions a metal-binding site. Substitution of Cys with seCys produces result in alteration to the protein structure and capacity due to the seCys being bigger, responsive and more effectively deprotonated than cysteine [79], as in the case of methionine sulfoxide reductase enzyme which lost its function as a result of the substitution of SeCys [19]. SeCys substitution mutilates the tertiary structure of protein because of its large diselenide bridge formation and modified redox potential affect enzyme kinetics [79]. Fe-S group proteins of chloroplast and mitochondrial electron transport chain [80] are inclined to SeCys substitution for instance as in the event of chloroplast NifS-like protein [81]. Fe-Se bunch are bigger in size and do not fit appropriately in apoproteins.

9.2 Selenium toxicity due to oxidative stress

A high dosage of selenium acts as a pro-oxidant and creates receptive oxygen species which cause oxidative stress in plants. Under selenium-induced stress, glutathione is diminished [82], except for Se-tolerant plants where raised level of glutathione is increased [83]. Previously studied plants such as *Arabidopsis* and *Viciafaba* have shown that reactive oxygen species accumulation under Se stress increased lipid peroxidation, cell mortality [20].

10. Selenium phytoremediation

Phytoremediation is a plant-based technology, which is eco-friendly, cheap and used in the treatment of contaminated soil and water resource [70, 84]. It does not reduce the fertility of the soil and this method of decontamination of the soil has been enhanced by the use of genetic engineering. Certain plants are suitable for phytoremediation, these plants are selected based of certain factors such as:

i. The plants must possess large biomass production capacity, volatilization and high accumulation of selenium

ii. It should be easily cultivated and harvested under different growing conditions

iii. The plants must have deep roots [70]

iv. They should be cheap to cultivate [85].

10.1 Methods of phytoremediation

There are various methods of phytoremediation, but phytovolatilization, rhizofiltration and phytoextraction are the most stable of selenium decontamination of soil and water bodies [70].

10.1.1 Photoextraction

Phytoextraction is the use of higher plants which are se-hyperaccumulators in the removal of se-contaminates from the soil [70, 86]. These se-hyperaccumulator plants grow on seleniferous soil and they are able to accumulate up to 15,000 mg/kg selenium. These plants are cultivated on the contaminated soil then after they have successfully removed the se in the soil they are disposed of. The main drawback of the method is that these se-hyperaccumulator plants grow slowly; this makes this strategy time consuming. They also have limited biomass production this leads to insufficient selenium decontamination for the soil [87].

10.1.2 Phytovolatilization

Phytovolatilization is the process of plants absorbing contaminants from the soil and releasing it to the atmosphere. Green plants are able to convert inorganic forms of selenium which are toxic to a less toxic organic selenocompounds [70]. This method is advantageous over phytoextraction, because it does not require the disposal of contaminated plant. The particular volatile selenium released by se-hyperaccumulator plants is dimethyldiselenide, while nonaccumulator plants release dimethylselenide from its leaves [88]. A more efficient method of phytoremediation is the combination of phytovolatilization and phytoextraction. This method increases the se-decontamination of soil by 2–3 times more than when carried out individually. Phytovolatilization method depends on certain factors such as the specie of plant, the microorganism in the rhizosphere, the selenium specie, temperature and so on [89].

10.1.3 Rhizofiltration

This strategy uses plant roots to decontaminate flowing water. It uses plant biomass to remove the contaminants as in the case of phytoextraction. Although they share same principle, rhizofiltration is used to decontaminant strictly water bodies and it involves the disposal of the root and shoot of the contaminated plant unlike phytoextraction which is only used to decontaminate the soil and involves the disposal of only the shoot of the contaminated plant.

10.1.4 Selenium biofortification

Selenium biofortification is a method used in the disposal of waste plants by decomposition [90, 91]. The selenium present in the plant is used to enrich the soil which aids in improvement of food quality [92]. Biofortification is an agricultural practice used in enriching food productions with different nutrients such as selenium in this case, with the purpose of increasing dietary intake by various biotechnological methods such as genetic engineering, plant breeding and manipulation of agronomic practices practices (Kieliszek and Blazejak, 2012) [93, 94]. Genetic engineering is a useful method of obtaining Se-biofortified food products, this is carried out by manipulation of selenium-related enzymes for uptake, evaporation and assimilation of selenium. Biofortification is cheap, safe and it also helps in carving out various nutrient deficiencies in diets [95–97]. Selenium biofortification is used to increases selenium contents of farm produces, this helps reduce selenium malnutrition among a population.

Author details

Abiodun Humphrey Adebayo*, Omolara Faith Yakubu and Osarobo Bakare-Akpata
Department of Biochemistry, College of Science and Technology, Covenant University, Ota, Ogun State, Nigeria

*Address all correspondence to: abiodun.adebayo@covenantuniversity.edu.ng

References

[1] Ekelung NGA, Danilov RA. The influence of selenium on photosynthesis and "light-enhanced dark respiration" (LEDR) in the flagellate *Euglena gracilis* after exposure to ultraviolet radiation. Aquatic Sciences. 2001;**63**:457-465

[2] Xue T, Hartikainen H, Piironen V. Antioxidative and growth-promoting effect of selenium in senescing lettuce. Plant and Soil. 2001;**237**:55-61

[3] Hartikainen H. Biogeochemistry of selenium and its impact on food chain quality and human health. Journal of Trace Elements in Medicine and Biology. 2005;**18**:309-318. DOI: 10.1016/j.jtemb.2005.02.009

[4] Funes-Collado V, Morell-Garcia A, Rubio R, Lopez-Sanchez J. Study of selenocompounds from selenium-enriched culture of edible sprouts. Food Chemistry. 2013;**141**:3738-3743

[5] Reilly C. Selenium in Food and Health. New York, NY, USA: Springer Science Media; 2006

[6] Patai S, Rappoport Z. The Chimestry of Organis Selenium and Tellurium Compounds. Vol. 1. New York, NY, USA: Willey; 1986

[7] Tinggi U. Essentiality and toxicity of selenium and its status in Australia: A review. Toxicology Letters. 2003;**137**:103-110

[8] Burk RF. Selenium in Biology and Human Health. New York, NY, USA: Springer-Verlag New York Inc.; 1994. p. 221

[9] Simonoff M, Simonoff G. Le sélénium et la vie. Paris, France: Masson; 1991. p. 242

[10] Fan RWM, The SJ, Hinton DE, Higashi RM. Selenium biotransformations into proteinaceous forms by foodweb organisms of selenium-laden drainage waters in California. Aquatic Toxicology. 2002;**57**:65-84

[11] Shardendu SN, Boulyga SF, Stengel E. Phytoremediation of selenium by two helophyte species in subsurface flow constructed wetland. Chemosphere. 2003;**50**(8):967-973

[12] White PJ, Brown PH. Plant nutrition for sustainable development and global health. Annals of Botany. 2010;**105**:1073-1080

[13] Sors TG, Ellis DR, Na GN, Lahner B, Lee S, Leustek T, et al. Analysis of sulfur and selenium assimilation in *Astragalus* plants with varying capacities to accumulate selenium. The Plant Journal. 2005;**42**:785-797. DOI: 10.1111/j.1365-313X.2005.02413.x

[14] Li HF, McGrath SP, Zhao FJ. Selenium uptake, translocation and speciation in wheat supplied with selenite or selenite. The New Phytologist. 2008;**178**:92-102. DOI: 10.1111/j.1469-8137.2007.02343.x

[15] Pilon-Smits EAH, Quinn CF. Selenium metabolism in plants. In: Hell R, Mendel R, editors. Cell Biology of Metal and Nutrients. Berlin: Springer; 2010. pp. 225-241

[16] Renkema H, Koopmans A, Kersbergen L, Kikkert J, Hale B, Berkelaar E. The effect of transpiration on selenium uptake and mobility in durum wheat and spring canola. Plant and Soil. 2012;**354**:239-250. DOI: 10.1007/s11104-011-1069-3

[17] Wallenberg M, Olim E, Hebert C, Björnstedt M, Fernandes AP. Seelnium compounds are substrates for glutaredoxins: A novel pathway for selenium metabolism and potential mechanism for selenium-mediated cytotoxicity. The Biochemical Journal. 2010;**429**:85-93

[18] Mroczek-Zdyrska M, Wójcik M. The influence of selenium on root growth and oxidative stress induced by lead in *Viciafaba* L. minor plants. Biological Trace Element Research. 2012;**147**:320-328

[19] Châtelain E, Satour P, Laugier E, LyVu B, Payet N, Rey P, et al. Evidence for participation of the methionine sulfoxide reductase repair system in plant seed longevity. Proceedings of the National Academy of Sciences of the United States of America. 2013;**110**:3633-3638

[20] Lehotai N, Kolbert Z, Peto A, Feigl G, Ördög A, Kumar D, et al. Selenite-induced hormonal and signaling mechanisms during root growth of *Arabidopsis thaliana* L. Journal of Experimental Botany. 2012;**63**:5677-5687

[21] Bisson M, Gay G, Guillard D, Ghillebaert F, Tack K. Le sélénium et ses composés. 2019. Available from: http://www.ineris.fr/substances/fr/substance/getDocument/3012 [Accessed: 28 September 2019]

[22] Gonzalez-Flores JN, Shetty SP, Dubey A, Copeland PR. The molecular biology of selenocysteine. Biomolecular Concepts. 2013;**4**(4):349-365. DOI: 10.1515/bmc-2013-0007

[23] Maroc L. Exposition professionnelle au sélénium et ses effets sur l'homme [Ph.D. thesis]. Paris, France: Université Paris 11 Chatenay; 1990

[24] Barceloux DG. Selenium. Journal of Toxicology. Clinical Toxicology. 1999;**37**:145-172

[25] Martens DA, Suarez DL. Selenium speciation of soil/sediment determined with sequential extractions and hydride generation atomic absorption spectrophotometry. Environmental Science & Technology. 1996;**31**:133-139

[26] Lebreton P, Salat O, Nicol JM. Un point sur le sélénium. Bulletin of Science and Technology. 1998:35-47

[27] Underwood EJ, Suttle NF. The Mineral Nutrition of Livestock. Cambridge, UK: CABI Publishing; 2004. p. 614

[28] Stadlober M, Sager M, Irgolic KJ. Effects of selenate supplemented fertilisation on the selenium level of cereals—Identification and quantification of selenium compounds by HPLC-ICP-MS. Food Chemistry. 2001;**73**:357-366

[29] Coughtrey PJ, Jackson D, Thorne MC. Selenium. In: Radionuclide Distribution and Transport in Terrestrial and Aquatic Ecosystems. Vol. 3. Rotterdam, The Netherlands: A. Balkema; 1983. 372p

[30] Neve J, Favier A. Selenium in medecine and biology. In: Proceedings of the Second International Congress on Trace Elements in Medecine and Biology, Avoriaz, France. Avoriaz, France, New York: Walter de Gruyer; 1988

[31] William G, Rambour S, Evrard CM. Physiologie des plantes. Bruxelles, Belgique: De boeck; 2003. 514p

[32] Fournier E. Bioaccumulation du sélénium et effets biologiques induits chez le bivalve filtreur corbicula fluminea. Prise en compte de l'activité ventilatoire, de la spéciation du sélénium et de la voie de contamination [Ph.D. thesis]. Bordeaux, France: Universite de Bordeaux 1; 2005

[33] Dumont E, Vanhaecke F, Cornelis R. Selenium speciation from food source to metabolites: A critical review. Analytical and Bioanalytical Chemistry. 2006;**385**:1304-1323

[34] Navarro-Alarcon M, Cabrera-Vique C. Selenium in food and the

human body: A review. Science of the Total Environment. 2008;**400**:115-141

[35] Whanger PD. Selenium and its relationship to cancer: An update. The British Journal of Nutrition. 2004;**91**:11-28

[36] Cabaraux JF, Dotreppe O, Hornick JL, Istasse L, Dufrasne I. Les oligo-éléments dans l'alimentation des ruminants: État des lieux, formes et efficacité des apports avec une attention particulière pour le sélénium, 2007. In: CRA-W-Fourrages Actualités, 12ème journée. 2007. pp. 28-36

[37] Suttle NF. Mineral Nutrition of Livestock. 4th ed. London, UK: MPG Books Group; 2010. p. 565

[38] Brigelius-Flohe R, Aumann KD, Blocker H, Gross G, Kiess M, Kloppel KD, et al. Phospholipid-hydroperoxide glutathione peroxidase. Genomic DNA, cDNA, and deduced amino acid sequence. The Journal of Biological Chemistry. 1994;**269**:7342-7348

[39] Ducros V, Favier A. Selenium metabolism. Endocrinología y Nutrición. 2004;**1**:19-28

[40] Meschy F. Nutrition minérale des ruminants; Editions Quae. France: Versaille; 2010. p. 208

[41] Chu FF, Doroshow JH, Esworthy RS. Expression characterization, and tissue distribution of a new cellular selenium-dependent glutathione peroxidase, GSHPx-GI. The Journal of Biological Chemistry. 1993;**268**:2571-2576

[42] Schwaab V, Faure J, Dufaure JP, Drevet JR. Gpx3: The plasma-type glutathione peroxidase is expressed under androgenic control in the mouse epididymis and vas deferens. Molecular Reproduction and Development. 1998;**51**:362-372

[43] Maiorino M, Scapin M, Ursini F, Biasolo M, Bosello V, Flohe L. Distinct promoters determine alternative transcription of gpx-4 into phospholipid-hydroperoxide glutathione peroxidase variants. The Journal of Biological Chemistry. 2003;**278**:34286-34290

[44] Brigelius-Flohe R, Maiorino M. Glutathione peroxidases. Biochimica et Biophysica Acta. 2013;**18**:3309

[45] Finch JM, Turner RJ. Effects of selenium and vitamin e on the immune responses of domestic animals. Research in Veterinary Science. 1996;**60**:97-106

[46] Ren F, Chen X, Hesketh J, Gan F, Huang K. Selenium promotes T-cell response to TCRstimulation and con a, but not PHA in primary porcine splenocytes. PLoS One. 2012;**7**:e35375

[47] Sordillo LM. Selenium-dependent regulation of oxidative stress and immunity in periparturient dairy cattle. Veterinary Medicine International. 2013;**2013**:e154045

[48] Broome CS, McArdle F, Kyle JAM, et al. An increase in selenium intake improves immune function and poliovirus handling in adults with marginal selenium status. The American Journal of Clinical Nutrition. 2004;**80**:154-162

[49] Davis CD, Tsuji PA, Milner JA. Selenoproteins and cancer prevention. Annual Review of Nutrition. 2012;**32**:73-95

[50] Koyama H, Mutakin, Abdulah R, Yamazaki C, Kameo S. Selenium supplementation trials for cancer prevention and the subsequent risk of type 2 diabetes mellitus. Nihon Eiseigaku Zasshi. 2013;**68**:1-10

[51] Maiorino M, Flohe L, Roveri A, Steinert P, Wissing JB, Ursini F. Selenium and reproduction. BioFactors. 1999;**10**:251-256

[52] Vendeland SC, Deagen JT, Butler JA, Whanger PD. Uptake of selenite, selenomethionine and selenate by brush border membrane vesicles isolated from rat small intestine. Biometals. 1994;**7**:305-312

[53] Oster O, Prellwitz W. The daily dietary selenium intake of west German adults. Biological Trace Element Research. 1989;**20**:1-14

[54] Kaur R, Kaur K. Effects of dietary selenium (Se) on morphology of testis and cauda epididymis in rats. Indian Journal of Physiology and Pharmacology. 2000;**44**:265-272

[55] Arechiga CF, Vázquez-Flores S, Ortiz O, Hernández-Cerón J, Porras A, McDowell LR, et al. Effect of injection of β-carotene or vitamin e and selenium on fertility of lactating dairy cows. Theriogenology. 1998;**50**:65-76

[56] Harrison JH, Russell Conrad H. Effect of dietary calcium on selenium absorption by the nonlactating dairy cow 1,2,3. Journal of Dairy Science. 1984;**67**:1860-1864

[57] Mistry HD, Pipkin FB, Redman CW, Poston L. Selenium in reproductive health. American Journal of Obstetrics and Gynecology. 2012;**206**:21-30

[58] Thomson CD, Robinson MF. Urinary and fecal excretions and absorption of a large supplement of selenium: Superiority of selenate over selenite. The American Journal of Clinical Nutrition. 1986;**44**:659-663

[59] Spears JW, Weiss WP. Role of antioxidants and trace elements in health and immunity of transition dairy cows. Veterinary Journal. 2008;**176**:70-76

[60] Kobayashi Y, Ogra Y, Ishiwata K, Takayama H, Aimi N, Suzuki KT. Selenosugars are key and urinary metabolites for selenium excretion within the required to low-toxic range. Proceedings of the National Academy of Sciences of the United States of America. 2002;**99**:15932-15936

[61] Bodnar M, Konieczka P, Namiesnik J. The properties, functions and uses of selenium compounds in living organisms. Journal of Environmental Science and Health, Part C. 2012;**30**:225-252

[62] Wu Z, Bañuelos GS, Lim ZQ, Lin Y, Yaun L, Yin X. Biofortifictaion and phytoremediation of selenium in China. Frontiers in Plant Science. 2015;**6**:136. DOI: 10.3389/fpls.2015.00136

[63] Missana T, Alonso U, Garcia-Gutierrez M. Experimental study and modelling of selenite sorption onto illite and smectite clays. Journal of Colloid and Interface Science. 2009;**334**:132-138. DOI: 10.1016/j.jcis.2009.02.059

[64] Kikkert J, Berkelaar E. Plant uptake and translocation of inorganic and organic forms of selenium. Archives of Environmental Contamination and Toxicology. 2013;**65**:458-465. DOI: 10.1007/s00244-013-9926-0

[65] Zhang Y, Pan G, Chen J, Hu Q. Uptake and transport of selenite by soybean seedlings of two genotypes. Plant and Soil. 2003;**253**:437-443. DOI: 10.1023/A:1024874529957

[66] White PJ, Bowen HC, Parmaguru P, Fritz M, Spracklen WP, Spiby RE, et al. Interactions between selenium and sulphur nutrition in Arabidopsis thaliana. Journal of Experimental Botany. 2004;**55**:1927-1937. DOI: 10.1093/jxb/erh192

[67] Harris J, Schneberg KA, Pilon-Smits EA. Sulfur–selenium–molybdenum interactions distinguish selenium hyperaccumulator *Stanleya pinnata* from non-hyperaccumulator

Brassica juncea (Brassicaceae). Planta. 2014;**239**:479-491. DOI: 10.1007/s00425-013-1996-8

[68] Mazej D, Osvald J, Stibilj V. Selenium species in leaves of chicory, dandelion, lamb's lettuce and parsley. Food Chemistry. 2008;**107**:75-83. DOI: 10.1016/j.foodchem.2007.07.036

[69] Gigolashvili T, Kopriva S. Transporters in plant sulphur metabolism. Frontiers in Plant Science. 2014;**5**:422

[70] Gupta M, Gupta S. An overview of selenium uptake, metabolism, and toxicity in plants. Frontiers in Plant Science. 2017;**7**:2074-2088. DOI: 10.3389/fpls.2016.02074

[71] Chu J, Yao X, Zhang Z. Responses of wheat seedlings to exogenous selenium supply under cold stress. Biological Trace Element Research. 2010;**136**:355-363. DOI: 10.1007/s12011-009-8542-3

[72] Pennanen A, Xue T, Hartikainen. Protective role of selenium in plant subjected to severe UV irradiation stress. Journal of Applied Botany and Food Quality. 2002;**76**:66-76

[73] Chen CC, Sung JM. Priming bitter gourd seeds with selenium solution enhances germinability and antioxidative responses under sub-optimal temperature. Physiologia Plantarum. 2001;**111**:9-16

[74] Germ M, Osvald J. Selenium treatment affected respiratory potential in Eruca sativa. Acta Agriculture Slovenica. 2005;**85**:329-335

[75] Aslam M, Harbit KB, Hufftaker RC. Comparative effects of selenite and selenate on nitrate assimilation in barley seedlings. Plant, Cell & Environment. 1990;**13**:773-782

[76] Van Hoewyk D. A tale of two toxicities: Malformed selenoproteins and oxidative stress both contribute to selenium stress in plants. Annals of Botany. 2013;**112**(6):965-972

[77] Trelease SF, Beath OA. Selenium: Its Geological Occurrence and its Biological Effects in Relation to Botany, Chemistry, Agriculture, Nutrition, and Medicine. Burlington, Vermont: The Champlain Printers; 1949

[78] Mengel K, Kirkby EA. Principles of Plant Nutrition. 4th ed. Worblaufen-Bern, Switzerland: International Potash Institute; 1987. ISBN No 3906535037

[79] Hondal RJ, Marino SM, Gladyshev VN. Selenocysteine in thiol/disulfide-like exchange reactions. Antioxidants & Redox Signaling. 2013;**18**(13):1675-1689. DOI: 10.1089/ars.2012.5013

[80] Balk J, Pilon M. Ancient and essential: The assembly of iron–Sulfur clusters in plants. Trends in Plant Science. 2011;**16**(4):218-226

[81] Pilon-smits EA, Garifullina GF, Abdel-ghany S, Kato S, Mihara H, Hale KL, et al. Characterization of a NifS-Like Chloroplast Protein from Arabidopsis. Implications for its role in Sulfur and Selenium Metabolism. Plant Physiology. 2002;**130**(3):1309-1318

[82] Hugouvieux V, Dutilleu C, Jourdain A, Reynaud F, Lopez V, Bourguignon J. Arabidopsis putative selenium-binding protein1 expression is tightly linked to cellular sulfur demand and can reduce sensitivity to stresses requiring glutathione for tolerance. Plant Physiology. 2009;**151**:768-781

[83] Grant K, Carey NM, Mendoza M, Schulze J, Pilon M, Pilon-Smits EAH, et al. Adenosine 5-phosphosulfate reductase (APR2) mutation in *Arabidopsis* implicates glutathione deficiency in selenate toxicity. Biochemical Journal. 2011;**438**:325-335

[84] Reis AR, El-Ramady H, Santos EF, Gratão PL, Schomburg L. The societal relevance of se for human and environmental health: Biofortification and phytoremediation. In: Pilon-Smits EAH, Winkel LHE, Lin Z-Q, editors. Selenium in Plants. Berlin: Springer International Publishing; 2017. pp. 231-255. ISBN No 978-3-319-56248-3

[85] Robinson BH, Bañuelos G, Conesa HM, Evangelou MWH, Schulin R. The phytomanagement of trace elements in soil. Critical Reviews in Plant Sciences. 2009;**28**:240-266

[86] Kumar NPBA, Dushenkov V, Motto H, Raskin I. Phytoextraction: the use of plants to remove heavy metals from soils. Environmental Science & Technology. 1995;**92**:1232-1238

[87] Chaney RL. Plant uptake of inorganic waste constituents. In: Parr JF, Marsh PD, Kla JM, editors. Land Treatment of Hazardous Wastes. Park Ridge: Noyes Data Corporation; 1983. pp. 50-76

[88] Evans CS, Asher CJ, Johnson CM. Isolation of dimethyl diselenide and other volatile selenium compounds from Astragalus Racemosus (Pursh.). Australian Journal of Biological Sciences. 1968;**21**:13-20

[89] Terry N, Zayed AM. Selenium volatilization by plants. In: Frankenberger WT, Benson S, editors. Selenium in the Environment. New York: Marcel Dekker, Inc.; 1994. pp. 343-369. ISBN No 0-8247-8993-8

[90] Liu Y, Li F, Yin XB, Lin ZQ. Plant-based biofortification: from phytoremediation to Se-enriched agriculture products. In: Sharmaand SK, Mudhoo A, editors. Green Chemistry for Environmental Sustainability. Boca Raton, FL: CRC Press. 2011. pp. 341-356

[91] Lin ZQ, Haddad S, Hong J, Morrissy J, Bañuelos GS, Zhang LY. Use of selenium-contaminated plants from phytoremediation for production of selenium-enriched edible mushrooms. In: Bañuelos GS, Lin ZQ, Yin XB, editors. Selenium in the Environment and Human Health. Boca Raton, FL: CRC Press. 2014. pp. 124-126

[92] Bañuelos GS, Arroyo I, Pickering IJ, Yang SI, Freeman JL. Selenium biofortification of broccoli and carrots grown in soil amended with Se-enriched hyperaccumulator Stanley apinnata. Food Chemistry. 2015;**166**:603-608

[93] Zhu YG, Pilon-Smits EAH, Zhao FJ, Williams PN, Meharg AA. Selenium in higher plants: understanding mechanisms for biofortification and phytoremediation. Trends in Plant Science. 2009;**19**:436-442

[94] Borrill P, Connorton JM, Balk J. Biofortification of wheat grain with iron and zinc:integrating novel genomic resources and knowledge from model crops. Frontiers in Plant Science. 2014;**5**:53-75

[95] Nestel P, Bouis HE, Meenakshi JV, Pfeiffer W. Biofortification of staple food crops. The Journal of Nutrition. 2006;**136**:1064-1067

[96] Mayer JE, Pfeiffer WH, Beyer P. Biofortified crops to alleviate micro nutrient malnutrition. Current Opinion in Plant Biology. 2008;**11**:166-170

[97] Zhao F-J, McGrath SP. Biofortification and Phytoremediation. Current Opinion in Plant Biology. 2009;**12**(3):373-380. DOI: 10.1016/j.pbi.2009.04.005

Chapter 4

Selenium Fertilization in Tropical Pastures

Letícia de Abreu Faria, Pedro Henrique de Cerqueira Luz and Adibe Luiz Abdalla

Abstract

Brazil is one of the largest meat producers. Meat along with other animal products have been responsible for its larger contribution as source of selenium (Se) for human. However, Se deficiency remains a concern because researches have indicated that this nutrient is found in low levels in Brazilian diet. Cattle in Brazil are fed basically from pasture, but there are strong evidences that soils contain low availability of Se; consequently plants and animals incorporate low Se levels. Pastures, Se fertilized, bring benefits to nutrition and health of animal consequently to humans already known in some countries. In contrast, Se fertilization on tropical weathered soils and tropical forages is little known. However, Se management as fertilizer in tropical environments requires researches involving field experiments, especially with animals, for establishing of safe and effective Se recommendations as fertilizer due to the Se toxicity potential and complexity in system of soil-plant-animal-human.

Keywords: agronomic biofortification, grazing animal nutrition, selenium fertilization, tropical forages, weathered soils

1. Introduction

Selenium (Se) benefits for health in human and animal have provided popularity for this chemical element. Selenium is a nutrient for animals since 1957 [1]; thus, it must be part of their diet. However, there are large agricultural areas containing low levels of Se in soil or it is present, but as chemical forms unavailable for plants, consequently these areas are producing vegetables or animal products with low Se contents.

In Brazil, there are some researches indicating that large agricultural areas are located in soils with low Se levels, that is, daily diet Se intake evaluated in people groups from São Paulo, Brazil, presented values below to estimated average requirement values, which shows deficiency of Se in the diets from this region [2]. A research at Rio Grande do Sul, Brazil, also resulted in marginal deficiency of Se in cattle [3]. Both data indicated the possibility of Brazilian soils contain low available Se levels.

The agronomic biofortification of food through field fertilization with Se could be a solution to provide this micronutrient for animals and humans through plants. Plants are able to absorb and incorporate Se to organic compounds as seleno-amino acids. Thus, inorganic Se is converted to organic Se compounds through the plants which can be easily absorbed by the human body and to be available where needed in the body [4]. In some countries, Se fertilization is well established, and it is annually done in New Zealand in which Se along with phosphorus fertilization is applied in pastures [5].

Selenium essentiality for plants is not convinced, but its availability for plants, as well as for the animals, could improve their performance and consequently the human health. For both animals and plants, this element acts as defense through its influence in glutathione peroxidase controlling oxygen reactive species from stress situations [6].

Large knowledge about specific rates, sources, Se dynamic in soils and plants, and even behavior of animal intake in pasture is required for a safe Se fertilization, to ensure food and environmental security. High Se levels available in soils can cause toxicity for plants and animals. Thus, for the beneficial of Se application as fertilizer commonly are required low rates, which raises concerns about risks of super dosages, what can be aggravated by complex dynamic of Se in soil and plants.

2. Selenium fertilization in tropical weathered soils

Se availability in soil is the first requirement for Se application as fertilizer. Most soils contain low Se levels, including in tropical environments, while the highest contents are found at arid areas characterized by the presence of accumulator plants [7].

Low Se levels were observed in the main eight soil types of Brazil (**Table 1**). These soils were collected at São Paulo state as well as in plant of *Urochloa decumbens* grown in them.

The contents up to 500 μg dm^{-3} characterized low Se soils [8], and confirming the relation between soil and plants, the samples of *Urochloa decumbens* comprised contents of 10.4–79.7 μg kg^{-1} Se in dry matter (unpublished data).

Besides the Se presence in soil, its availability for plants depends on oxidation state. Selenium is chemically similar to sulfur, but it occurs naturally in four oxidation states, −2 (selenide), 0 (elemental Se), +4 (selenite), and +6 (selenate) [9]; however, sodium selenate is the source recommended for Se fertilizations due its

		Depth	pH	O.M.	P	S	K	Ca	Mg	H+Al	Al	CTC	SB	V	m	B	Cu	Fe	Mn	Zn	Se
Soils	Localization	(cm)	$CaCl_2$	g.dm^{-3}	mg.dm^{-3}		mmolc.dm^{-3}							%		mg.dm^{-3}					(μg.kg^{-1})
Arenic Hapludult	22º38,366'S	0-20	4,4	14	4	6	1,1	6	3	30	3	40	10	25	21	0,15	1,3	73	2,1	0,7	67,9
	47º49,852'W	20-40	4,5	15	3	8	0,4	7	3	24	1	34	10	30	12	0,13	1,1	60	2,9	0,3	84,6
Typic Fluvaquent	22º42,407'S	0-20	5,5	27	5	12	5,2	41	6	27	-	79	52	66	-	0,19	7,5	28	28,5	1,9	220,2
	47º37,438'W	20-40	5,2	21	4	13	3,5	24	5	29	1	62	33	53	4	0,16	6,2	15	21,1	0,9	219,4
Acroxic Dystrochept	22°38,404'S	0-20	5,5	31	7	17	1,7	58	8	35	-	103	68	66	-	0,24	1,9	75	50	3,5	108
	47°49,024'W	20-40	5,3	24	5	12	1,2	55	9	35	-	100	65	65	-	0,16	1,7	53	49,5	2,1	138,4
Typic Haplodoll	22º04,961'S	0-20	4,7	18	3	8	0,5	8	4	31	1	43	12	29	7	0,08	1,3	43	0,9	0,5	159,6
	47º34,736'W	20-40	4,4	19	3	7	0,3	4	3	41	4	48	7	15	36	0,16	1,4	51	0,6	0,9	142,1
Acroxic Dystrochept	22º 15,054'S	0-20	4	65	6	6	0,4	3	2	240	5	245	5	2	49	0,36	1,9	76	0,6	0,7	80,3
	47º52,044'W	20-40	3,9	32	4	6	0,3	5	3	105	1	113	8	7	12	0,21	2,1	34	0,2	0,3	72,9
Typic Quartzipsamment	21º56,630'S	0-20	5,5	24	4	8	1,1	39	6	23	-	69	46	67	-	0,16	1,8	39	2,9	3,4	77,6
	47º28,506'W	20-40	5,4	20	3	10	1,0	29	5	20	-	55	35	64	-	0,13	1,6	22	1,1	0,6	101,7
Typic Haplodux	21º 57,768'S	0-20	5,3	28	33	9	1,2	40	8	29	1	78	49	63	2	0,16	8,3	40	7,1	5,7	197,1
	47º26,866'W	20-40	5,3	21	9	8	0,9	21	4	27	1	53	26	49	4	0,10	6,3	22	6	1,8	158,8
Rhodic Haplodux	21º 35,278'S	0-20	4,2	12	2	5	2,6	7	4	128	1	142	14	10	7	0,10	0,8	11	2,6	0,2	97,8
	48º26,054'W	20-40	4,2	12	2	5	2,3	6	3	151	2	162	11	7	17	0,07	0,6	10	1,3	0,1	80,5

Table 1.
Chemical parameters of fertility in tropical soils and selenium levels in sampled soils at São Paulo state, Brazil.

high solubility. Unlike selenate, the mechanism of selenite uptake by plants remains unclear [10].

In weathered soils, there are low nutrient levels; high contents of Fe, Al, and Mn; and high acidity, according to soils analyzed (**Table 1**). Thus, it is necessary to know the Se dynamic in these soils. A profile of weathered soils analyzed from São Paulo observed low Se levels in soils with higher sand contents (**Figure 1**) that could indicate leaching potential.

A study of Se adsorption and desorption in soils from Cerrado, Brazil, verified low values of distribution coefficient in soils; thus, Se tended to be more in solution than in the solid phase, and in the most weathered soils, with higher clay and Al and Fe oxide contents, there are the highest affinity for Se, while in sandy and loamy soils, Se tends to be less adsorbed and can therefore be taken up by plants or easily leached, damaging the ecosystem [11].

The low natural levels of Se in soils and its absence in fertilization to crops explain the low contents in food from vegetables [12] and consequently in Brazilian diet, except for northern areas [2]. Although, in a study of hemodialysis patients from north and southeast of Brazil, both patient groups presented low Se plasma levels when compared to recommended values; independently of the region, all patients presented Se deficiency [13].

This information is an alert for necessity to Se fertilization in Brazil. It was incentivized in the 1980s, but its requirement was unsuccessful, while in some countries, it is well stablished already. Applications of Se in areas of low Se bioavailability have been an option with good results to supply this element to plants and, consequently to animals, improving animal performance and nutritional quality of food produced as milk and meat [11], even in environments with no deficiency symptoms [14].

However, toxic potential of selenium requires caution as fertilizer due its complex dynamic nature in soil and plants. Selenium as selenate (SeO_4^{2-}) is commonly found in alkaline and oxygen soils under high redox conditions (pe + pH > 15), and this oxidation state is predominantly absorbed by plants, while under low and milder redox conditions, species as selenite and selenide predominate [9].

Selenium fertilization must be controlled through safe doses and soil monitoring due to the possibilities of the Se dynamics in soils. Selenate can be easily absorbed by plants; thus, the doses for fertilizations must be carefully calculated, but also it could be leached with possibilities of water contamination.

Low selenate doses required for fertilization is a challenge due to concern for homogeneous application. According to the Selenium-Tellurium Development

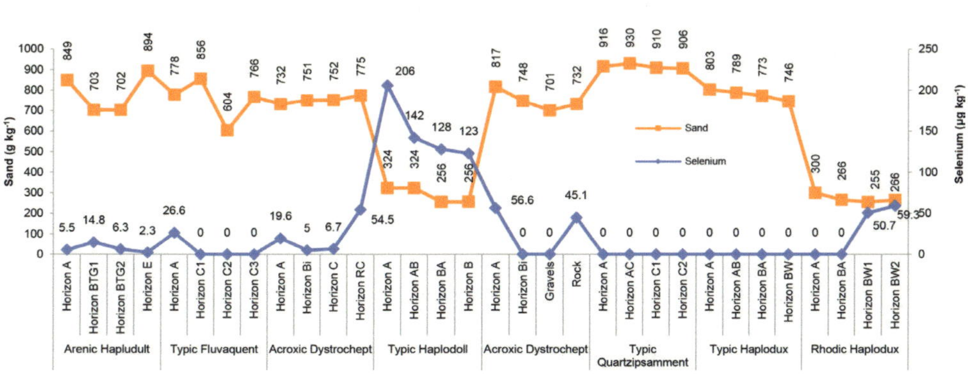

Figure 1.
Selenium levels in tropical soil profiles and, respectively, sand contents in sampled soils at São Paulo state, Brazil (unpublished data).

Association, the best way for Se application is along with other nutrients [15]. There are positive effects in Se application along with phosphorus fertilization [13].

Some technologies involving Se application along with macronutrients as coating became a technique to easy and high quality of application. Urea coated with a mix of boric acid (0.4% B), copper sulfate (0.14% Cu), and sodium selenate applied to *Urochloa brizantha* carried out in pots with weathered tropical soil resulted to desirable enrichment of plant with rate of 34.5 g ha^{-1} Se [16].

Seed pelletization seems to be a promissory tool to increase Se content in plants. Beneficial effects were observed in the evaluation of seed pelletization with increasing selenite doses on three ryegrass cultivars; however, the authors recommended it to be evaluated under field conditions in Se-deficient soils [17].

Another technology is the slow release of Se fertilizer as Selcote Ultra; however, its application in rates of up 20 g ha^{-1} Se on an Ultisol soil of Puerto Rico did not increase in the foliage Se concentrations of Guinea grass pastures [18]. According to the authors, the soil and plant interrelationships may be affecting the foliage of Se absorption potential requiring that the effects need future consideration in terms of Se movement in tropical soils.

The establishment of effective and safe rates in tropical environmental still is required and unknown, regardless of the technical method applied to plant enrichment on Se. High Se availability in agriculture soils can cause toxicity to crops, but it is still more concerning if a crop shows accumulator character, i.e., if a crop has capacity to absorb high levels of selenium with no symptoms of toxicity; this could increase the possibilities to cause toxicity for animals or human.

Forage plants are classified as passive accumulators due its ability to contain 10–30 mg kg^{-1} of Se in dry matter; however, high-quantity animal intake of Se through dry matter intake could induce intoxication [19]. Although the research with *Urochloa brizantha* grown in a weathered tropical soil containing 1.8–4.6 mg dm^{-3} Se resulted in high levels of Se uptake affecting biomass production, regardless of the soil type, plants showed high levels of Se in leaves [20].

Depending of the soil, excess of Se can be in unavailable forms to plant uptake over time. Soil influence in dynamic and availability of Se was observed by isolation of Selenium rates applied followed by comparing among soils, Arenic Hapludult, Rhodic Hapludox and Typic Hapludoll, which verified differents Selenium content remaining in soil, respectively, 22, 11, and 37 μg dm^{-3} and 55, 4, and 38 μg dm^{-3} after *Urochloa brizantha* and *Stylosanthes capitata* cultivated, respectively (unpublished data).

Besides the uptake ability among plants, the difference among soils was evident. This element in soil can be fixed along with iron, complexed in organic matter, or it can be in many oxidation states depending on the pH, oxygen, and microbial activity [7]. These data confirmed different soil capacities to Se adsorption, but it can turn available for plants in pH changes, as frequently occurs with limestone application, a common practice from tropical agriculture areas.

High contents of selenate in soil, for natural or anthropogenic action, can be establish or reestablish for agricultural or livestock, avoiding the poisoning risk of plants, animals, and humans [20]; in these cases, the areas must be isolated for remediation. The use of plants to clean up contaminated soils is a technique known as phytoremediation that offers a less expensive alternative to stripping pollutants directly from the soil [21].

The management of high Se soil also can include sulfur source application. The similarity between Se and S indicates the competition for sulfate transporters of the root plasma membrane [7]. Sulfur application at 600 kg ha^{-1} as soluble sources such as ammonium sulfate and ferric sulfate reduced damages in productivity and Se uptake by Urochloa grass, while the lower solubility of calcium sulfate resulted in lower effectiveness in reducing Se uptake [20]. Plants used for phytoremediation can be used for mixture in diets of animals or used as composting to application in low Se areas.

3. Selenium fertilization for tropical grazing

The geochemistry of livestock-producing areas should be well understood to mitigate selenium-related disorders in animals [22]. In China, there are areas with selenosis occurrences; in contrast about 70% of China shows Selenium deficiency, as well as there are about 76% countries located in Se-deficient regions where the Se daily intake level is less than recommended [23].

Brazil is a country who owns the largest livestock in pastures; however, the majority of pastures are in marginal areas, under soils of low fertility. Brazil is one of the largest meat producers. In Brazilian food, the highest Se concentrations have been found to animal origin products, while vegetable food showed lower values [12]; however, some studies showed selenium deficiencies also in cattle, except in north areas [24, 25].

Cattle deficiencies have a likely relation between low Se in forage grass or supplements [26]. Thus, Se deficiency keeps as a concern, since this nutrient has been found in low levels in Brazilian diet.

Selenium fertilization could be a solution, but it is not realized in Brazil. Fertilizers containing selenium could support diet supplementation of grazing animals or animals feed with conserved forages (hay and silage), mainly considering Brazilian pasture areas with more than 160 million of hectares [27].

Selenium fertilization in pastures could increase its presence in animal and human diet besides its benefits to animal productivity. Feeding weaned beef calves for 7 weeks with alfalfa hay containing up to 3.26 mg kg^{-1} Se in dry matter harvested in fertilized fields with Se resulted to increasing whole-blood Se concentrations and body weights depending upon the Se application rate [28].

Grazing Se fertilizer has been shown to be more effective and safe treatment than animal dosing [13]. Selenium intake by ruminants in organic form, especially as selenoamino acids, needs to be release from proteins until through inorganic forms to be metabolized while Se in milk is from organic Se supply or converted by ruminal microorganisms [29].

Experiment with tropical grass has been shown large differences between Se contents in leaves and stem + sheath. *Urochloa brizantha* had desirable increased Se in leaves (0.4 mg kg^{-1} Se in DM) even in smaller evaluated dose of 10 g ha^{-1} Se up to 30 days, while the proportion of stem + sheath (0.1 mg kg^{-1} Se in DM) however for safe and required concentration in dry matter for cattle along with prolonged effect of two cuts, the authors recommended Se fertilization of 34.5 g ha^{-1} [16].

The fast answer of Se fertilization in tropical pastures can be positive to allow the low and safe doses of Se application along with nitrogen rates during rainy season, including for intercropped pastures with legumes. The concerning of intercropped pastures is explained by legume higher ability to produce protein, which can be apparently favorable to Se absorption.

Thus, this fact probably will require lower Se doses for fertilization in intercropped pastures, including caution with palatability and proportion of legumes in pastures. For example, in three different weathered soils carried out in pots, fertilization of 20 g ha^{-1} Se showed desirable concentrations to legume *Stylosanthes capitata* (136 μg kg^{-1} Se in DM), while it was insufficient to grass *Urochloa brizantha* (49.1 μg kg^{-1} Se in DM); however, it showed negative influence in content of crude protein from legume, probably by influence in nitrogen biological fixation (unpublished data).

High Se contents in leaves from tropical grass and in legumes comprise another fact to be analyzed for Se fertilization rate establishment although with few data. Usually, grass leaves and legumes are preference fractions for cattle according to its intake selective behavior, mainly in tropical pastures due to high accumulation of stem portions and its low digestibility. Selenium is one of few elements absorbed by plants in enough quantities which enable to intoxicate domestic animals [30].

Evaluating in vitro degradability of two cuts of *Urochloa brizantha* produced using fertilization rates of 0, 10, 20, 40, and 80, and 160 g ha^{-1} Se verified effects on gas production by truly degraded organic matter, amounts of acetic by propionic acids, some short-chain fatty acids and ammonia amounts [31]. The authors suggested that high levels of Se in forage can affect negatively ruminal microorganism activities but indicates positive effects of Se fertilization in 20 and 80 g ha^{-1} on chemical composition, in vitro degradability, short-chain fatty acids, and gas production.

In superior plants dual effects can be exerted by Se; at low concentrations it acted as an antioxidant, inhibiting lipid peroxidation, whereas at higher concentrations, it was a prooxidant [6]. Applied doses of selenate above 71 g ha^{-1} Se exceed maximum recommendation of 5 mg kg^{-1} Se in dry matter of the leaves of *Urochloa brizantha* to avoid toxicity problems in cattle [32] however with no damages in dry matter production [16]. Thus, even the plant is aparently normal, is necessary attempt that the recommendation for nutritional requirements of animals are values between 0.1 to 0.3 mg kg^{-1} in the dry matter required by cattle [20].

The effect of high Se diet concentration (60–70 μg kg BW^{-1} d^{-1}) provided from wheat to steers indicated the negative effects of Se level used in this study on productive performance of feedlot which were not expected [25]. Low forage digestibility can contribute to low Se effects in animals, mainly in tropical forages, while high Se content in forage can reduce its digestibility. Undegraded residues from in vitro incubation contained 25–66% of Se from *Urochloa brizantha* Se enriched [31].

4. Final considerations

Selenium fertilization in tropical low Se soils, as Brazil agricultural areas, is an emergent necessity for animal and human health, also could be beneficial for plants. Thus, Se fertilization in pastures is an alternative to collaborate for animal supplementation and human nutritional demands.

Generally, Se quantities required as fertilizer are low, and there are already available technologies to application but is an extremely necessary soil monitoring.

Nevertheless, more researches in tropical environments is required to establishment of Se rates, plants, and animal answers and reduces or even neutralizes toxicity risks, even though the benefits already are known.

Author details

Letícia de Abreu Faria[1*], Pedro Henrique de Cerqueira Luz[2] and Adibe Luiz Abdalla[3]

1 Universidade Federal Rural da Amazônia, Paragominas, Pará, Brazil

2 Faculdade de Zootecnia e Engenharia de Alimentos of Universidade de São Paulo, Pirassununga, São Paulo, Brazil

3 Centro de Energia Nuclear na Agricultura of Universidade de São Paulo, Piracicaba, São Paulo, Brazil

*Address all correspondence to: leticiadeabreufaria@gmail.com

References

[1] Sillanpää M, Jansson H. Status of cadmium, lead, cobalt and selenium in soils and plants of thirty countries. In: FAO Soils Bull. n. 65. 1992. Disponível em: http://books.google.com.br/books/about/Status_of_Cadmium_Lead_Cobalt_and_Seleni.html?id=Uiy2O_y7ZSIC&redir_esc=y [Acesso em: 17 Jan. 2013]

[2] Maihara VA, Gonzaga IB, Silva VL, Fávaro DIT, Vasconcellos MBA, Cozzolino SMF. Daily dietary selenium intake of selected Brazilian population groups. Journal of Radioanalytical and Nuclear Chemistry. 2004;**259**:465-468. DOI: 10.1023/B:JRNC.0000020919.58559.dd

[3] Valle SF, González FD, Rocha D, Scalzilli HB, Campo R, Larosa VL. Mineral deficiencies in beef cattle from southern Brazil. Brazilian Journal of Veterinary Research and Animal Science. 2003;**40**:47-53

[4] Agency for Toxic Substances and Disease Registry (ATSDR). Toxicological Profile for Selenium. Atlanta, GA: U.S. Department of Health and Human Services, Public Health Service; 2003

[5] Watkinson JH. Prevention of selenium deficiency in grazing animals by annual top dressing of pasture with sodium selenate. New Zealand Veterinary Journal. 1983;**31**:78-85

[6] Hartikainen H, Xue T, Piironen V. Selenium as an anti-oxidant and prooxidant in ryegrass. Plant and Soil. 2000;**225**:193-200

[7] Malavolta E. Selênio. In: Elementos de nutrição mineral de plantas. São Paulo: Agronômica Ceres; 1980. pp. 211-212

[8] Millar KR. Selenium. In: Grace ND, editor. The Mineral Requirements of Grazing Ruminants. New Zealand: New Zealand Society of Animal Production; 1983. pp. 38-47

[9] Hawkesford MJ, Kok LJ. Sulfur in Plants and Ecological Perspective. The Netherlands: Springer; 2007. pp. 225-252

[10] Li HF, Mcgrath SP, Zhao FJ. Selenium uptake, translocation and speciation in wheat supplied with selenate or selenite. New Phytologist. 2008;**178**:92-102. DOI: 10.1111/j.1469-8137.2007.02343.x

[11] Abreu LB, Carvalho GS, Curi N, Guilherme LRG, Marques JJGSM. Sorção de Se em solos do bioma cerrado. Revista Brasileira de Ciência do Solo. 2011;**35**:1995-2003

[12] Ferreira KS, Gomes JC, Bellato CR, Jordão CP. Concentrações de selênio em alimentos consumidos no Brasil. Pan American Journal of Public Health. 2002;**11**(3):172-177

[13] Stockler-Pinto MB, Malm O, Azevedo SRG, Farage NE, Dorneles PR, Cozzolino SMF, et al. Selenium plasma levels in hemodialysis patients: Comparison between North and Southeast of Brazil. Jornal Brasileiro de Nefrologia. 2014;**36**(4):490-495

[14] Whelan BR, Barrow NJ, Peter DW. Selenium fertilizers for pastures grazed by sheep. II.* Wool and liveweight responses to selenium. Australian Journal Agriculture Research. 1994;**45**:877-887

[15] Selenium-Tellurium Development Association. Se & Te. Disponível em: http://www.stda.net/ [Acesso em: 21 ago. 2007]

[16] Faria LA, Machado MC, Karp FHS, Kamogawa MY, Abdalla AL. Brachiaria enrichment with selenium-coated urea. Ciencia Rural. 2018;**48**(06):e20170630

[17] Cartes P, Gianfreda L, Paredes C, Mora ML. Selenium uptake and its antioxidant role in ryegrass cultivars as affected by selenite seed palletization. Journal of Soil Science and Plant Nutrition. 2011;**11**(4):1-14

[18] Santana RR, Vazquez A, McDowell LR, Wilkinson NS, Macchiavelli R. Selenium fertilization of Guinea grass pastures in Central Puerto Rico. Communications in Soil Science and Plant Analysis. 2006;**37**:621-625

[19] Oliveira KD, França TN, Nogueira VA, Peixoto PV. Enfermidades associadas à intoxicação por selênio em animais. Pesquisa Veterinaria Brasileira. 2007;**27**:125-136

[20] Faria LA, Luz PHC, Macedo FB, Tonetti PS, Ferraz MR, Mazza JA, et al. Brachiaria in selenium-contaminated soil under sulphur source application. Revista Brasileira de Ciência do Solo. 2015;**39**:1814-1820

[21] Sirko A, Gotor C. Molecular links between metals in the environment and plant sulfur metabolism. In: Hawkesford MJ, Kok LJ, editors. Sulfur in Plants and Ecological Perspective. The Netherlands: Springer; 2007. pp. 225-252

[22] Khanal DR, Knight AP. Selenium: Its role in livestock health and productivity. The Journal of Agriculture and Environment. 2010;**11**:101-106

[23] Wu Z, Bañuelos GS, Lin ZQ, Liu Y, Yuan L, Yin X, et al. Biofortification and phytoremediation of selenium in China. Frontiers in Plant Science. 2015;**6**:1-8. DOI: 10.3389/fpls.2015.00136

[24] Moraes SS, Tokarnia CH, Döbereiner J. Deficiências e desequilíbrios de microelementos em bovinos e ovinos em algumas regiões do Brasil. Pesquisa Veterinaria Brasileira. 1999;**19**:19-33

[25] Soto-Navarro SA, Lawler TL, Taylor JB, Reynolds LP, Reed JJ, Finley JW, et al. Effect of high-selenium wheat on visceral organ mass, and intestinal cellularity and vascularity in finishing beef steers. Journal of Animal Science. 2004;**82**(6):1788-1793

[26] Gil et al. Selenium in bovine plasma, soil and forage measured by neutron activation analysis. Arquivo Brasileiro de Medicina Veterinária e Zootecnia. 2004;**56**(2):264-266

[27] Instituto Brasileiro de Geografia e Estatística (IBGE). Sistema IBGE de Recuperação Automática. Brasília: Distrito federal; 2006. [online]. Available from: http://www.sidra.ibge.gov.br/bda/pecua/default.asp?z=t&o=24&i= [Accessed: 17 Jan 2013]

[28] Hall JA, Saun RJV, Nichols T, Mosher W, Pirelli G. Comparison of selenium status in sheep after short-term exposure to high-selenium-fertilized forage or mineral supplement. Small Ruminant Research. 2009;**82**:40-45

[29] Gierus M. Fontes orgânicas e inorgânicas de selênio na nutrição de vacas leiteiras: Digestão, absorção, metabolismo e exigências. Ciência Rural. 2007;**37**(4):1212-1220

[30] Dhillon KS, Dhillon SK. Distribution and management of seleniferous soils. Advances in Agronomy. 2003;**79**:119-184

[31] Faria LA, Karp FHS, Righeto PP, Abdalla Filho AL, Lucas RC, Machado MC, et al. Nutritional quality and organic matter degradability of Brachiaria spp. agronomically biofortified with selenium. Journal of Animal Physiology and Animal Nutrition. 2018:**102**(6):1464-1471. DOI: 10.1111/jpn.12971

[32] National Research Council—NRC. Mineral Tolerance of Animals. Washington: National Academy Press; 2005

Chapter 5

Chromatographic Analysis of Selenium Species

Aleksandra Sentkowska

Abstract

Selenium is an important element in environmental and living organisms that is being essential in very narrow concentration range, while deficiency or toxicity occurs outside this range. However, its toxicity depends not only on its dose but also on its chemical form. In environmental samples, selenium can exist in inorganic forms (as elemental selenium, metal selenides, selenite, or selenate anions) and as organic species with direct C-Se bonds (methylated compounds, selenoaminoacids, and selenoproteins). Thus, the development of reliable techniques to study the speciation of selenium in environmental samples is necessary. The main purpose of this chapter is to provide an update on the recent literature concerning the strategies for selenium speciation in environmental samples. Liquid chromatography coupled with sensitive detector is a commonly used technique for selenium separation. Gas chromatography can also be applied for such purpose; however derivatization step is usually required before analysis. Direct determination of selenium species at the concentration levels present in natural samples is very often difficult or even impossible. For this, several preconcentration/separation procedures for selenium have been proposed, including coprecipitation, extraction into an organic solvent, or application of solid sorbents.

Keywords: selenium, speciation, chromatography, gas chromatography, liquid chromatography

1. Introduction

Selenium and its several species have been demonstrated to be essential for living organisms [1, 2]. It plays a key role in many important metabolic pathways such as thyroid hormone metabolism and antioxidant defense systems [3]. For these reasons, it should be present in human diet. The range between the deficiency and toxicity of selenium is very narrow. The nutritionally required daily uptake of selenium is 55 μg; however some studies suggest that it should be 100 μg [4]. It is estimated that the diets of 1 billion people might lack sufficiency for their well-being [5, 6]. The toxic dose of selenium is very much dependent on its chemical form, with different toxicity for organic and inorganic forms [7]. Selenoaminoacids are principal dietary forms of selenium, and selenomethionine (SeMet) is derived from plants, while selenocysteine (SeCys) from animals [8, 9].

Selenium naturally exists in many different inorganic (elemental selenium, selenide, selenite, and selenate ions) and organic forms (methylated compounds, selenoaminoacids, and selenoproteins). Inorganic selenium, present in water and soil, can be easily transformed into volatile compounds by plants and fungi.

	IUPAC name	Abbreviation	Chemical formula
Air	Diethylselenide	DESe	CH_3-CH_2-Se-CH_2-CH_3
Soil and water	Selenite	Se(IV)	SeO_3^{2-}
	Selenate	Se(VI)	SeO_4^{2-}
Plants	Dimethylselenide	DMSe	CH_3-Se-CH_3
	Dimethyldiselenide	DMDSe	CH_3-Se-Se-CH_3
	Dimethylseleniumsulfide	DMSSe	CH_3-Se-S-CH_3
	Selenocysteine	SeCys	HSe-CH_2CH(NH_2)-COOH
	Selenocystine	$SeCys_2$	HCOOH-CH(NH_2)CH_2-Se-Se-CH_2CH(NH_2)-COOH
	Se-Methylselenocysteine	SeMC	CH_3Se-CH_2CH(NH_2)-COOH
	Selenomethionine	SeMet	CH_3Se-$CH_2$$CH_2$CH($NH_2$)-COOH
	Se-Methylselenomethionine	MeSeMet	$(CH_2)_2Se^+$-CH_2-CH_2-CH(NH_2)-COOH
	γ-Glutamyl-Se-methylselenocysteine	γ-Glu-SeMC	H_2N$CH_2$$CH_2$-CO-NGCH(COOH)$CH_2$-Se-$CH_3$
	Selenocystathionine	SeCysth	HCOOH-CH(NH_2)-CH_2-CH_2-Se-CH_2-CH(NH_2)COOH
	Selenocystamine	SeCystm	H_2N-CH_2-CH_2-Se-Se-CH_2-CH_2-NH_2
	Se-Adenosylselenohomcysteine	AdoSeHcy	NH_2-CH(COOH)-CH_2-CH_2-Se-CH_2-$C_4H_5O_3C_5N_4$-NH_2

Table 1.
Principal selenium species present in environmental samples.

Organic species of selenium form covalent C-Se bonds. SeCys is included into selenoproteins and participates in redox reactions. The metabolic pathway of selenium in the human body is complicated [10]. In general it can be divided into three groups including reduction of inorganic species by glutathione (GHS) to selenide, cleavage reaction of organic species by β-lyase; utilization according to the UGA codon leading the synthesis of selenoproteins; and finally excretion after being metabolized to methylated species.

Apart from the importance of the selenium in living organisms, this element is also spread throughout the environment. Sulfur-containing minerals are natural sources of selenium, but it is also produced by combustion of fossil fuels. It should be noticed that selenium is used in electronic industries and agriculture as a component of fertilizers.

Several analytical procedures for determination of selenium at low concentration levels in environmental samples have been proposed and recently reviewed [11, 12]. For selenium speciation analysis, the coupling of chromatographic techniques such as gas chromatography (GC) or high-performance liquid chromatography (HPLC) with a highly sensitive and selective detector is very useful [13]. Even though GC exhibits high efficiency and simplicity, HPLC has the ability of dealing with non-volatile compounds, extending the range of application and avoiding a derivatization step. This chapter will focus on the recent progress in the application of HPLC in different modes for selenium speciation analysis in water, soil, and plants. Sample pretreatment procedures will be also considered. The principal selenium species present in environmental samples are summarized in **Table 1**.

2. Sample preparation

Sample preparation step is crucial in every analysis where analytes are present at very low concentration levels. In the speciation analysis, there is also another difficulty that should be overcome. An important requirement for reliable speciation is to retain the concentration and structure of the original chemical forms in the sample. In general, aquatic samples such as rain, ground- or surface water, tap and drinking water, seawater, and soil solutions do not require any pretreatment procedures other than filtration through 0.45 μm filter.

Extraction of selenium species from the solid samples with the highest recovery is quite challenging. According to Peachey et al. [14], selection of the extraction method, which provides high extraction efficiency while preventing the integrity of selenium species, is essential for the accurate measurement of its species. The most used method can be divided into three main groups:

- Extraction using aqueous solution (water, water-methanol, and buffers).
- Acid or alkali hydrolysis (hydrochloric acid and methanesulfonic acid).
- Enzymatic hydrolysis (proteinase, protease, and mixture of enzymes).

Since selenoaminoacids are water-soluble, extraction with hot water is extensively used [14–17]. However, the efficiency of water extraction from yeast was only 10% [17]. To release bounded selenoaminoacids, enzymatic or acidic hydrolysis was necessary [18–20]. The addition of methanesulfonic acid was used for selenomethionine extraction from yeast when heated under reflux [21]. Casiot et al. [22] reported that extraction of selenium species from yeast with water and ethanol led only to 10–20% recoveries of selenium and not allowed to extract selenomethionine. The addition of pectinolytic enzymes released additional 20% of selenomethionine, while the addition of dodecyl sulfate solution allowed solubilization of a selenoprotein that accounted for 30% of total selenium. On the other hand, using tetramethylammonium hydroxide solubilized the sample completely, but the extracted selenium species were entirely degraded to selenomethionine and inorganic selenium [22]. It should be noticed that this type of extraction strictly depends on the choice of an enzyme, pH of the extraction solution, as well as temperature and time of the extraction. The most commonly used enzymes for this purpose are proteinase K or proteolytic enzymes (protease XIV), which were used for water-insoluble selenium fraction in many complicated matrices [23]. To reduce the extraction time, also ultrasonic hydrolysis can be used as the breakdown of Se-containing proteins (peptide bonds) into selenoaminoacids occurs [24]. The in vitro digestion with gastric juice was also used for selenium extraction from fish samples [25]. This procedure allowed SeMet determination, but the whole process takes few hours.

The sequential procedure developed by Chassaigne et al. [26] consisted of three steps: first, Tris-HCl buffer was used for extraction of water-soluble fractions, then also Tris-HCl with the addition of SDS for solubilization of protein fraction, and, finally, concentrated HNO_3 was used for dissolving of the remaining solid residue. A similar three-step procedure was used for extraction of selenium from mushrooms with 89% extraction efficiency [27]. It should be noticed that sequential treatment was also applied for dietary supplements [28, 29].

Cuderman et al. [30] examined different extraction media to identify selenium species in buckwheat. The optimal extraction efficiencies were obtained by hydrolysis with HCl, followed by breaking the cells with liquid nitrogen and then enzymatic hydrolysis with protease XIV. Ammonium acetate [31], sodium hydroxide [32], and enzymatic hydrolysis with pronase E [33] have been proposed for extraction of selenium species from green onion leaves.

3. Total selenium determination

Determination of the total selenium content is still the first step of its analysis. This procedure requires that organic forms must be transferred into inorganic selenium that is usually achieved with digestion using strong mineral acids or UV irradiation after addition of hydrogen peroxide [12, 34]. For this purpose, fluorometry, electrochemical detectors, atomic absorption spectrometry (AAS), atomic fluorescence spectrometry (AFS), mass spectrometry (MS), and neutron activation analysis have been used (**Table 2**).

Hydride generation coupled to AAS or AFS detectors is specific for Se(IV) determination, where these species are selectively reduced to volatile SeH_2, usually by sodium tetrahydroborate in hydrochloric medium. This technique can be applied for the determination of total inorganic selenium (e.g., sum of selenite and selenate) after quantitative reduction. The content of Se(VI) is then obtained by the difference between two determinations. This technique can be fully automatic by connection with flow injection analysis system. The advantage of such system is minimum sample and reagent consumption as well as short time of single run.

Stripeikis et al. [44] determined selenite and selenate in drinking water using fully automatic online separation/pre-concentration system coupled to electrochemical atomic spectrometry. Preconcentration of both selenium forms was carried out onto a microcolumn packed with an anionic resin (Dowex 1X8) that was placed in the robotic arm of the autosampling device. Selenite and selenate were then sequentially eluted with HCl at concentration of 0.1 and 4 mol L^{-1},

Technique	Matrix	Detection limit	Ref.
UV-Vis	Water, soil	0.012 mg L^{-1}	[34]
Fluorometry	Water	0.35 ng	[35]
X-ray fluorescence	Water	0.032 ng mL^{-1}	[36]
HGAAS	Water	2 ng L^{-1}	[37]
ETAAS	Seafood	0.16 μg g^{-1}	[38]
ICP-MS	Soil	3–29 ng L^{-1}	[39]
Adsorptive stripping voltammetry	Water	0.06 ng mL^{-1}	[40]
Differential-pulse cathodic stripping voltammetry	Food supplements	0.2 ng mL^{-1}	[41]
Differential-pulse anodic stripping voltammetry		0.06 ng mL^{-1}	[42]
Instrumental neutron activation analysis	Libyan food	26–90 μg L^{-1}	[43]

Table 2.
Analytical methods for the determination of total selenium.

respectively. The interference of large amounts of chloride ions during selenium atomization was prevented by using iridium as a permanent chemical modifier.

Kocot et al. [36] proposed a dispersive micro-solid-phase extraction with graphene as a solid adsorbent and ammonium pyrrolidine dithiocarbamate as a chelating agent for Se(IV) in analysis of inorganic selenium by the energy-dispersive X-ray fluorescence spectrometry; the concentration of Se(VI) was calculated as the difference between the concentration of selenite after and before reduction of selenate.

Due to the high selectivity and sensitivity, wide linear range, as well as multielement and multi-isotope detection, inductively coupled plasma mass spectrometry (ICP-MS) is a great tool for selenium analysis. However, some difficulties can be found when conventional ICP-MS is used for this purpose. To avoid the spectra interferences with 80Se isotope (the largest natural abundance of 49.6%), 82Se and 77Se are often monitored. The use of collision/reaction cell, operating with hydrogen gas, lowers argon dimer interferences. This technique was applied by Reyes et al. to determine the selenium in biological reference materials and serum samples [45] and yeast [46].

4. Separation techniques in selenium speciation analysis

The occurrence of selenium at very low concentration levels as well as the dependence of its toxicity on the form in which it occurs resulted in the need for reliable analytical procedures for identification and quantification of its species. The coupling of separation of selenium species with sensitive detection has become a powerful technique for Se speciation. Liquid chromatography operating in different modes is the most used analytical technique for this purpose; however electrophoresis and gas chromatography also were used in selenium analysis. That is why these two methods will be also described in this chapter, which will mainly focus on liquid chromatography.

4.1 Gas chromatography

Selenium species can be divided into volatile and nonvolatile compounds. The first group of compounds can be directly analyzed using gas chromatography, for example, dimethylselenide was determined in human breath after ingestion of Se-enriched selenite [47]. For other nonvolatile compounds, derivatization is required.

In the literature many derivatization procedures can be found. Pelaez et al. [48] tested two methods of derivatization of selenomethionine and selenomethionine in selenium supplements. The first method consisted of esterification of the carboxylic acid group using propan-2-ol and then acylation of amino group. The second described procedure used ethanol-ethyl chloroformated for one step derivatization. Then detection using ICP-MS was performed.

In general, many types of detectors were coupled with GC for speciation analysis of selenium. Yang et al. [49] successfully used ICP detection for the determination of selenium in yeast. Organoselenium species in plants were determined using GC-MIP-AES detection [50]. Because of the requirement of the derivatization, GC is not so widely used in selenium speciation analysis in comparison to liquid chromatography.

4.2 Capillary electrophoresis

Due to the high resolving power, capillary electrophoresis (CE) has a potential to be used in speciation analysis as an alternative or complementary technique to HPLC. CE coupled with ICP-MS was used in the analysis of selenium in yeast extract after SEC separation [51]. The obtained limit of detection for organic and inorganic species was in range 7–18 $\mu g\ L^{-1}$. The main difficulty in such separation can be big sizes of selenopolypeptides resulting in their slow migration. In such case, the predigestion of selenoprotein fraction is recommended.

It should be highlighted that gel electrophoresis offers better resolution than liquid chromatography in the analysis of high molecular weight selenoproteins, which was used in such analysis by Chassaigne et al. [26], and in Se-enriched yeast analysis as well by Chery et al. in blood sample analysis [52].

4.3 High-performance liquid chromatography

4.3.1 Size-exclusion chromatography

Separation in size-exclusion chromatography (SEC) strictly depends on the size of separated analytes. That is why this technique is widely used as a preliminary step to sample purification or to separate selenoproteins from the matrix. Ayouni [53] observed high molecular weight selenium compounds in the extract of dietary supplements after separation by SEC. Conjunction of SEC with ICP-MS was used for analysis of the products of enzymatic digestion of selenoproteins fraction [54, 55]. Size-exclusion chromatography was also used in selenium analysis in human plasma [56] and extracts from rats' internal organs [57].

4.3.2 Ion-exchange chromatography

Anion-exchange chromatography has been mainly employed in selenium speciation analysis [58–63]. Mobile phases used in this mode usually contain small content of organic modifier (e.g., 2–5% methanol) and buffered salt solution (e.g., acetate, phosphate, and citrate). During the separation process, the equilibria between the charged solute ion and the oppositely charged surface of the stationary phase are established. The separation is achieved based on the differences of the strength of such interaction between analytes. Anion-exchange chromatography was used in the selenium speciation analysis in garlic, sunflowers, and radish sprouts [64]. In addition to well-known compounds like SeMet or MeSeCys, several unidentified signals were obtained. The application of high-resolution mass spectrometry enabled identification of additional seleno-compounds as inorganic metabolites, such as deamino-hydroxy-seleno-homolanthionine, *N*-acetylcysteine-selenomethionine, methylseleno-pentose-hexose, methylselenoglutathione, 2,3-dihydroxy-propionyl selenocysteine-cysteine, methyltio-selenoglutathione, 2,3-dihydroxy propionyl-seleno-lanthionine, and two Se-containing compounds with proposed molecular formula $C_{10}H_{18}N_2O_6Se$ and $C_{10}H_{13}N_5O_3Se$.

Cation-exchange chromatography was used to analyze selenium-enriched yeast in a human adsorption study [65]. As a mobile phase, pyridinium formate buffer with 3% of methanol was used. This method was suitable for separation of organic selenium species, however not suitable to separate selenite and selenate.

4.3.3 Reversed-phase chromatography

Both, simple reversed-phase [66–68] and ion-pair (IP) reversed-phase chromatography [15, 28, 69], are widely used for analysis of ionic and neutral selenium species. The mobile phases are aqueous with small amount of polar organic solvent (usually methanol or acetonitrile). Because of their hydrophilicity, selenoamino-acids are not retained onto typical reversed-phase columns. The use of ion-pairing reagents as mobile phase additives allows their separation. The ion-pairing reagent is usually an alkyl sulfonate, an alkyl sulfate, or an alkylammonium salt. Its non-polar chain interacts with hydrophobic stationary phase (e.g., C8 or C18), while ionizable group is neutralized by oppositely charged analyte. Hexanesulfonic acid has been used as anion-pairing reagent in the speciation analysis of selenium in Brazil nuts, using C8 column for separation [70]. Obtaining separation was satisfied for organic compounds but poor for selenite and selenate. For separation of organic and inorganic forms of selenium, tetrabutylammonium acetate was proposed [71]. Also mixed ion-pairing reagents (butanesulfonic and tetramethylammonium hydroxide) were also used to simultaneously separate inorganic and organic species with satisfactory separation efficiencies [72].

New mobile phase additives are still developed, for example, room temperature ionic liquids [73]. Their mechanism of action is based on bilayer formulation onto stationary phase. It gives the possibility of additional interactions between the analyte and the bed, which significantly affects the retention and shape of the recorded signals. The effects of several imidazolium chlorides on the separation of selenium species mixture was described in details [74]. In all cases, SeMeCys was the first species eluted indicating its weak retention in the column, while the retention time of Se(VI) was increasing with the increase of alkyl chain.

4.3.4 Hydrophilic interaction liquid chromatography

Hydrophilic interaction liquid chromatography (HILIC) is a complementary technique to reversed-phase mode. The separation mechanism is mainly based on the partition of the analyte between the thin water layer adsorbed onto the stationary surface and the eluent, which contain high content of organic solvent usually acetonitrile. It is known that also other interactions such as hydrogen bonding dipole-dipole interactions and electrostatic forces may play an important role in the retention mechanism in HILIC [75]. The governed retention mechanism strictly depends on the type of used stationary phase and the buffer conditions (content of organic solvent, concentration of salt, and pH). TSKgel Amide-80 stationary phase with covalently bound carbamoyl groups is frequently used in the analysis of selenium in HILIC mode [76, 77]. According to the characterization of amide stationary phase, the achieved separation is not pH dependent [78]. Also zwitterionic and silica stationary phases have been also used in HILIC separation of selenium [79, 80]. It should be highlighted that in selenium separation in HILIC mode methanol is recommended instead of acetonitrile as a main component of mobile phase [79] which is shown in **Figure 1**.

The use of methanol enhances peak intensity, improve the separation of SeMet and SeMeSeCys, and shorten time of the single run. The best separation conditions

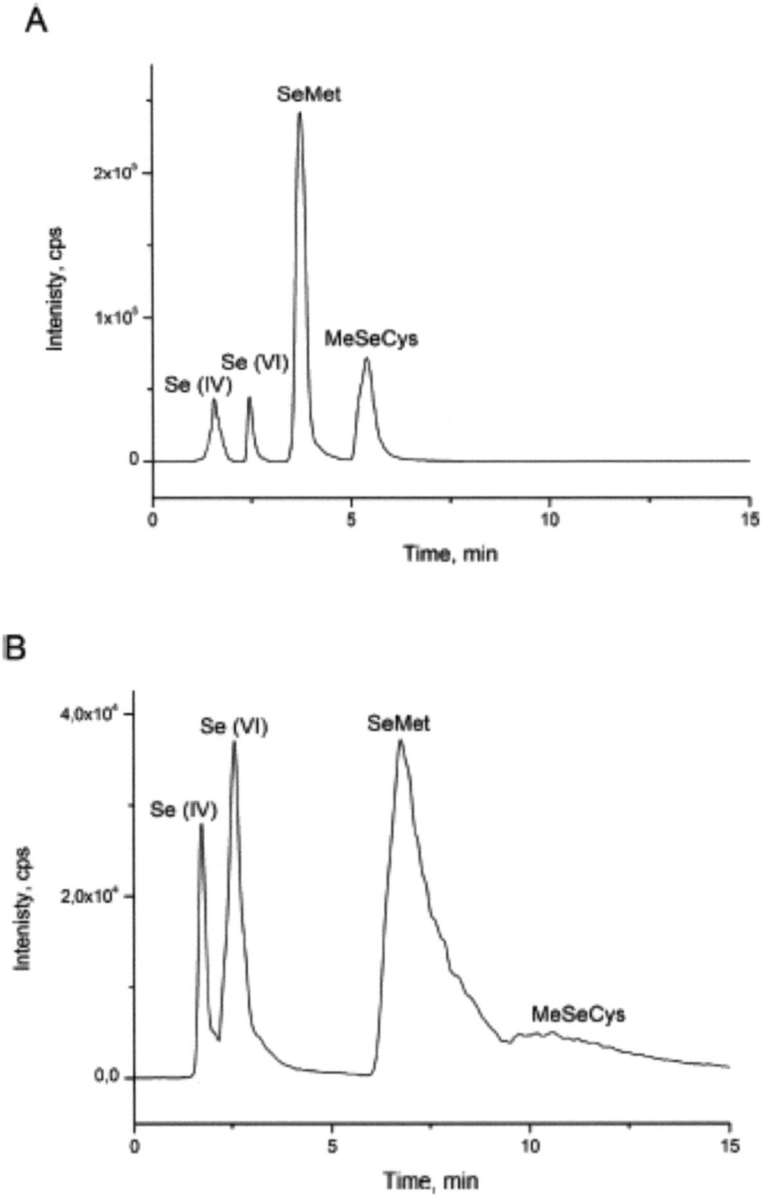

Figure 1.
The chromatogram of selenium compounds obtained on silica column using (A) MeOH and (B) ACN in the mobile phase [79]. Reprinted with permission from Elsevier.

have been obtained for silica column and mobile phase consisted of 85% of methanol and 8 mM of ammonium acetate. Using the zwitterionic column (ZIC-HILIC) instead of silica stationary phase resulted in recording of very asymmetric peaks.

The potential of two orthogonal chromatographic modes—RP and HILIC—was examined in the analysis of onion leaf extracts [79]. Higher separation efficiency (mainly for inorganic selenium species) and shorter retention times were obtained when HILIC mode was used (**Figure 2**).

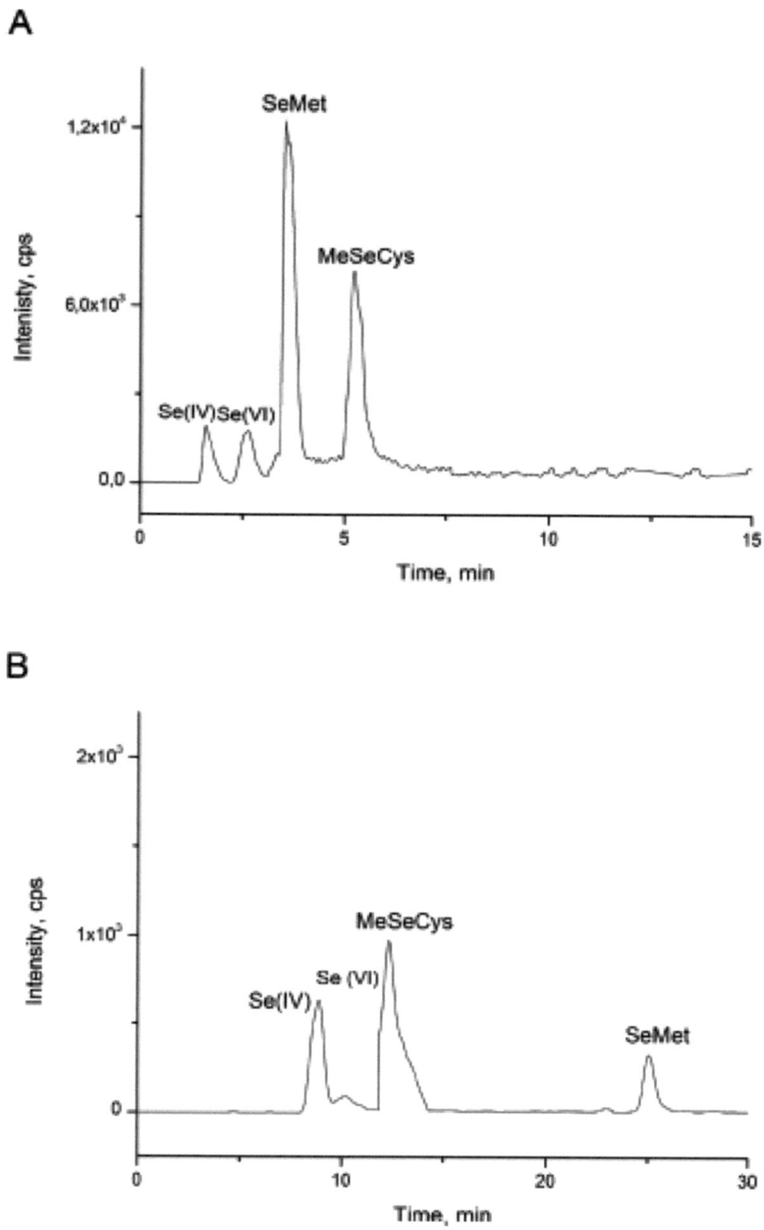

Figure 2.
The chromatographic separation of selenium species present in onion leaf extracts in (A) HILIC mode and (B) RP mode. (A) Atlantis HILIC (silica) column, mobile phase: 85% MeOH and 8 mM CH_3COONH_4, pH 7. (B) Luna C8 column, mobile phase: 99.5% HCOOH and 0.5% ACN [79]. Reprinted with permission from Elsevier.

5. Conclusions

The determination of selenium is of great importance from the point of view of understanding its metabolism and its potential benefits for human health. Due to the variety of selenium species and low concentrations in which it occurs in environmental samples, liquid chromatography coupled to ICP-MS is the most powerful method for speciation analysis of selenium. However, other techniques, for example, HILIC-MS/MS, have also been successfully applied in the speciation analysis of selenium. It should be highlighted that for such analysis all the analytical steps like selection of separation and detection method and the optimization of separation parameters (eluent type and composition, mobile phase additives) should be optimized to avoid the coelution of selenium species.

The recent application of liquid chromatography for selenium separation in plants (vegetables samples) is shown in **Table 3**. For sure new chromatographic strategies will be described in the literature in the nearest future.

Sample (determined selenium species)	Mobile phase	Stationary phase	Detection	Ref.
Reversed-phase chromatography				
Green leafy vegetables ($SeCys_2$, SeMet)	50% MeOH + 1.5% HCOOH	C18 (250 × 4.6 mm, 5 μm)	ICP-MS	[69]
Ion-pair reversed-phase chromatography				
Garlic (MeSeCys, SeMet, Se(IV), Se(VI))	Gradient phosphate buffer pH 6.0, 5% MeOH, 0.1% [C6mim]Cl	Agilent Zorbax SB-C8 (150 × 4.6 mm)	HG-AFS	[58]
Anion-exchange chromatography				
Onion (MeSeCys, SeMet, Se(IV), g-glu-MeSeCys, Se(VI))	Gradient acetate buffer, pH 4.7	Hamilton PRP-X100 (250 × 4.1 mm, 10 mm)	ICP-MS	[69]
Cation-exchange chromatography				
Carrot (inorganic Se, SeMet, g-glu-MeSeCys)	Gradient: ammonium formate, pH 3.0, 5% MeOH	Chrompack IonoSpher 5C (150 × 2.0 mm, 5 mm)	ICP-MS	[75]
Hydrophilic liquid chromatography				
Onion leaves (Se(IV), Se(VI), MeSeCys, SeMet)	Gradient: MeOH and ammonium acetate, pH 7	Atlantis HILIC (silica) (100 × 2.1 mm, 3.0 mm)	MS/MS	[81]

Table 3.
Examples of the recent HPLC application for selenium speciation in plant materials.

Conflict of interest

The author declares no conflict of interest.

Appendices and nomenclature

AAS	atomic absorption spectrometry
AFS	atomic fluorescence spectrometry

ACN	acetonitrile
C8	octylsilane column
CE	capillary electrophoresis
CH_3COONH_4	ammonium acetate
ETAAS	electrothermal atomic absorption
GC	gas chromatography
HCOOH	formic acid
HCl	hydrochloric acid
HGAAS	hydride generation atomic absorption spectrometric
HILIC	hydrophilic interaction liquid chromatography
HNO_3	nitric acid
HPLC	high-performance liquid chromatography
ICP-MS	inductively coupled plasma-mass spectrometry
ILs	ionic liquids
IP	ion pair
MeOH	methanol
MeSeMet	Se-methylselenomethionine
RP	reverse phase
Se(IV)	selenite
Se(VI)	selenate
SeCys	selenocysteine
SEC	size-exclusion chromatography
SeMet	selenomethionine

Author details

Aleksandra Sentkowska
Heavy Ion Laboratory, University of Warsaw, Warsaw, Poland

*Address all correspondence to: sentkowska@slcj.uw.edu.pl

References

[1] Tan LC, Nancharaiach YV, van Hullebusch ED, Lens PNL. Selenium: Environmental significance, pollution and biological treatment technologies. Biotechnology Advances. 2016;**34**:886-907. DOI: 10.1016/j.biotechadv.2016.05.005

[2] Weekeley CM, Harris HH. Which form is that? The importance of selenium speciation and metabolism in the prevention and treatment of disease. Chemical Society Reviews. 2013;**42**:8870-8894. DOI: 10.1039/c3cs60272a

[3] Zimmerm MT, Bayse CA, Ramoutar RR, Brumaghim JL. Sulfur and selenium antioxidants: Challenging radicals scavenging mechanism and developing structure-activity relationship based on metal binding. Journal of Inorganic Biochemistry. 2015;**145**:30-40. DOI: 10.1016/j.jinorgbio.2014.12.020

[4] Clark LC, Combs GF, Turnbull BW, Slate EH, Chalker DK, Chow J, et al. Effects of selenium supplementation for cancer prevention in patients with carcinoma of the skin. A randomized controlled trial. Nutritional Prevention of Cancer Study Group. Journal of the American Medical Association. 1996;**276**:1957-1963

[5] Joy EJM, Ander EL, Young SD, Black CR, Watts MJ, Chilimba ADC, et al. Dietary mineral supplies in Africa. Physiologia Plantarum. 2014;**31**:208-229. DOI: 10.1111/ppl.12144

[6] Stoffaneller R, Morse NL. A review of dietary selenium intake and selenium status in Europe and the Middle East. Nutrients. 2015;**7**:1494-1537. DOI: 10.3390/nu7031494

[7] Forchhammer K, Bock A. Biologie und biochemie des elements selen. Naturwissenschaften. 1991;**78**:497-504. DOI: 10.1007/BF01131397

[8] Olsen OE, Novacek EJ, Whitehead EI, Palmer IS. Investigation of selenium in wheat. Phytochemistry. 1970;**9**:1181-1188. DOI: 10.1016/S0031-9422(00)85306-6

[9] Hawkens WC, Wilhelmes EC, Tapper AL. Abundance and tissue distribution of selenocysteine-containing proteins in the rat. Journal of Inorganic Biochemistry. 1985;**23**:77-92. DOI: 10.1016/0162-0134(85)83011-7

[10] Wastney ME, Combs GF, Canfield WK, Taylor PR, Patterson KY, Hill DA, et al. A human model of selenium that integrates metabolism from selenite and selenomethionine. Japanese Journal of Nutrition. 2011;**141**:708-717. DOI: 10.3945/jn.110.129049

[11] Kumkrong P, LeBlanc K, Mercier PHJ, Mester Z. Selenium analysis in waters. Part 1: Regulation and standard methods. Science of the Total Environment. 2018;**640-641**:1611-1634. DOI: 10.1016/j.scitotenv.2018.05.392

[12] LeBlanc K, Kumkrong P, Mercier PHJ, Mester Z. Selenium analysis in waters. Part 2: Speciation methods. Science of the Total Environment. 2018;**640-641**:1635-1651. DOI: 10.1016/j.scitotenv.2018.05.394

[13] Marcinkowska M, Barałkiewicz D. Multielemental speciation analysis by advanced hyphenated technique—HPLC/ICP-MS: A review. Talanta. 2016;**161**:1777-1204. DOI: 10.1016/j.talanta.2016.08.034

[14] Peachey P, McCarthy N, Goenaga-Infante H. Acceleration of enzymatic hydrolysis of protein-bound selenium by focused microwave energy. Journal of Analytical Atomic Spectrometry. 2008;**23**:487-492. DOI: 10.1039/B717854A

[15] Kubacha KM, Hanley T, Mantha M, Wiklson RA, Falconer TM, Kassa Z,

et al. Evaluation of selenium in dietary supplements using elemental speciacion. Food Chemistry. 2017;**218**:313-320. DOI: 10.1016/j.foodchem.2016.08.086

[16] Tsai Y, Lin C, Hsu I, Sun Y. Sequential photocatalyst-assisted digestion and vapor generation device coupled with anion exchange chromatography and inductively coupled plasma mass spectrometry for speciation analysis of selenium species in biological samples. Analytica Chimica Acta. 2014;**806**:165-171. DOI: 10.1016/j.aca.2013.11.008

[17] Kotrebai M, Bird SM, Tyson JF, Block E, Uden PC. Characterization of selenium species in biological extracts by enhanced ion-pair liquid chromatography with inductively coupled plasma-mass spectrometry and by referenced electrospray ionization-mass spectrometry. Spectrochimica Acta B. 1999;**54**:1573-1591. DOI: 10.1016/S0584-8547(99)00086-5

[18] Kotrebai M, Birringer M, Tyson JF, Block E, Uden PC. Selenium speciation in enriched and natural samples by HPLC-ICP-MS and HPLC-ESI-MS with perfluorinated carboxylic acid ion-pairing agents. Analyst. 2000;**125**:71-78. DOI: 10.1039/A906320J

[19] Kotrebai M, Tyson JF, Block E, Uden PC. High-performance liquid chromatography of selenium compounds utilizing perfluorinated carboxylic acid ion-pairing agents and inductively coupled plasma and electrospray ionization mass spectrometric detection. Journal of Chromatography. A. 2000;**866**:51-63. DOI: 10.1016/S0021-9673(99)01060-2

[20] McSheehy S, Yang W, Pannier F, Szpunar J, Łobiński R, Auder J, et al. Speciation analysis of selenium in garlic by two-dimensional high-performance liquid chromatography with parallel inductively coupled plasma mass spectrometric and electrospray tandem mass spectrometric detection. Analytica Chimica Acta. 2000;**421**:147-153. DOI: 10.1016/S0003-2670(00)01039-4

[21] Wróbel K, Wróbel K, Caruso JA. Selenium speciation in low molecular weight fraction of Se-enriched yeast by HPLC-ICP-MS: Detection of selenoadenosylmethionine. Journal of Analytical Atomic Spectrometry. 2002;**17**:1048-1054. DOI: 10.1039/B200920J

[22] Casiot C, Szpunar J, Łobiński R, Potin-Gautier M. Sample preparation and HPLC separation approaches to speciation analysis of selenium in yeast by ICP-MS. Journal of Analytical Atomic Spectrometry. 1999;**14**:645-650

[23] Huerta VD, Sanchez MLF, Sanz-Medel A. Quantitative selenium speciation in cod muscle by isotope dilution ICP-MS with reaction cell: Comparison of different reported extraction procedures. Journal of Analytical Atomic Spectrometry. 2004;**19**:644-648. DOI: 10.1039/b313826g

[24] Capelo JL, Fernandez C, Pedras B, Santos P, Gonzales P, Vaz C. Trends in selenium determination/speciation by hyphenated techniques based on AAS or AFS. Talanta. 2006;**68**:1442-1447. DOI: 10.1016/j.talanta.2005.08.027

[25] Cabanero AI, Madrid Y, Camara C. Selenium and mercury bioaccessibility in fish samples: An in vitro digestion method. Analytica Chimica Acta. 2004;**526**:51-61. DOI: 10.1016/j.aca.2004.09.039

[26] Chassaigne H, Chery CC, Bordin G, Ropdriguez AR. Development of new analytical methods for selenium speciation in selenium-enriched yeast material. Journal of Chromatography. A. 2001;**976**:409-422. DOI: 10.1016/S0021-9673(02)00945-7

[27] Dernovics M, Stefanka Z, Fodor P. Improving selenium extraction by sequential enzymatic processes for Se-speciation of selenium-enriched *Agaricus bisporus*. Analytical and Bioanalytical Chemistry. 2002;**372**:473-480. DOI: 10.1007/s00216-001-1215-5

[28] Hsieh Y, Jiang S. Determination of selenium compounds in food supplements using reversed-phase liquid chromatography-inductively coupled plasma mass spectrometry. Microchemical Journal. 2013;**110**:1-7. DOI: 10.1016/j.microc.2013.01.009

[29] Wang W, Chen Z. Extraction of selenium species in pharmaceutical tablets using enzymatic and chemical methods. Microchimica Acta. 2009;**165**:167-172. DOI: 10,1007/s00604-008-0115-1

[30] Cuderman P, Ozbolt L, Krefl I, Stibilij V. Extraction of Se species in buckwheat sprouts grown from seeds soaked in various Se solutions. Food Chemistry. 2010;**123**:941-948. DOI: 10.1016/j.foodchem.2010.04.063

[31] Tie M, Li B, Sun T, Guan W, Liang Y, Li H. HPLC-ICP-MS speciation of selenium in Se-cultivated *Flammulina velutipes*. Arabian Journal of Chemistry. 2017 (in press). DOI: 10.1061/j.arabje.2017.05.012

[32] Kapolna E, Fodor P. Speciation analysis of selenium enriched green onions (*Allium fistulosum*) by HPLC-ICP-MS. Microchemical Journal. 2006;**84**:52-62. DOI: 10.1016/j.microc.2006.04.014

[33] Kapolna E, Shah M, Caruso JA, Fodor P. Selenium speciation studies in Se-enriched chives (*Allium schoenoprasum*) by HPLC-ICP-MS. Food Chemistry. 2007;**101**:1398-1406. DOI: 10.1016/j.foodchem.2006.03.048

[34] Ramachandran Kumar GS. Modified spectrophotometric method for the determination of selenium in environmental and mineral mixtures using 2,3-diaminonaphthalene. Talanta. 1996;**43**:1711-1714. DOI: 10.1016/0039-9140(96)01947-9

[35] Wang D, Alfthan G, Aro A. Determination of total selenium and dissolved selenium species in natural waters by fluorometry. Environmental Science and Technology. 1994;**28**:383-387. DOI: 10.1021/es00052a007

[36] Kocot K, Leardi R, Walczak B, Sitko R. Determination and speciation of trace and ultratrace selenium ions by energy-dispersive X-ray fluorescence spectrometry using graphene as solid adsorbent in dispersive micro-solid phase extraction. Talanta. 2015;**134**:360-365. DOI: 10.1016/j.talanta.2014.11.036

[37] Örnemark U, Pettersson J, Olin A. Determination of total selenium in water by atomic-absorption spectrometry after hydride generation and preconcentration in a cold trap system. Talanta. 1992;**39**:1089-1096. DOI: 10.1016/0039-9140(92)80206-S

[38] Méndez H, Alava F, Lavilla I, Bendicho C. Ultrasonic extraction combined with fast furnace analysis as an improved methodology for total selenium determination in seafood by electrothermal-atomic absorption spectrometry. Analytica Chimica Acta. 2002;**452**:217-222. DOI: 10.1016/S0003-2670(01)01475-1

[39] Tolu J, Hecho IL, Bueno M, Thiry Y, Potin-Gautier M. Selenium speciation analysis at trace level in solis. Analytica Chimica Acta. 2011;**684**:126-133. DOI: 10.1016/j.aca.2010.10.044

[40] Ashournia M, Aliakbar A. Determination of Se(IV) in natural waters by adsorptive stripping voltammetry of 5-nitropiazselenol. Journal of Hazardous Materials.

2010;**174**:788-794. DOI: 10.1016/j.jhazmat.2009.09.121

[41] Holak W, Specchio JJ. Determination of selenium in food supplements by differential-pulse cathodic stripping voltammetry in the presence of added copper. Analyst. 1994;**119**:2179-2182. DOI: 10.1039/AN9941902179

[42] Ramadan AA, Mandil H, Shikh-Debes A. Differential pulse anodic stripping voltammetric determination of selenium (IV) at a gold electrode modified with 3, 3′-diaminobenzidine. 4HCl-Nafion. International Journal of Pharmacy and Pharmaceutical Sciences. 2014;**6**:148-153. DOI: 10.24297/jac.v15i2.7818

[43] Elghawi U, Al-Sadeq AA, Bejey MM, Alamin MB. Determination of selenium in Libyan food items using pseudocyclic instrumental neutron activation analysis. Biological Trace Element Research. 2005;**107**:61-72. DOI: 10.1385/BTER:107:1:061

[44] Stripeikis J, Pedro J, Bonivardi A, Tudino M. Determination of selenite and selenite in drinking water: A fully automatic on-line separation/pre-concentration system coupled to electrochemical atomic spectrometry with permanent chemical modifiers. Analytica Chimica Acta. 2004;**502**:99-105. DOI: 10.1016/j.aca.2003.09.022

[45] Reyes LH, Gayon JMM, Alonso JIG, Sanz-Medel A. Determination of selenium in biological materials by isotope dilution analysis with an octopole reaction system ICP-MS. Journal of Analytical Atomic Spectrometry. 2003;**18**:11-16. DOI: 10.1039/B209213A

[46] Reyes LH, Marchante-Gayon JM, Alonso JIG, Sanz-Medel A. Application of isotope dilution analysis for evaluation of extraction conditions in the determination of total selenium and selenomethionine in yeast-based nutritional supplements. Journal of Agricultural and Food Chemistry. 2006;**54**:1557-1563. DOI: 10.1021/jf0523768

[47] Kremer D, Ilgen G, Feldman J. GC-ICP-MS determination of dimethylselenide in human breath after ingestion of 77 Se-enriched selenite: Monitoring of in-vivo methylation of selenium. Analytical and Bioanalytical Chemistry. 2005;**383**:509-515. DOI: 10.1007/s00216-005-0001-1

[48] Pelaez MV, Bayon MM, Alonso JIG, Sanz-Medel A. A comparison of different derivatisation approaches for the determination of selenomethionine by GC-ICP-MS. Journal of Analytical Atomic Spectrometry. 2000;**15**:1217-1222. DOI: 10.1039/B000505N

[49] Yang L, Mester Z, Stugeon RE. Determination of methionine and selenomethionine in yeast by specie-specific isotope dilution GC/MS. Analytical Chemistry. 2004;**75**:5149-5156. DOI: 10.1021/ac049475p

[50] Kahakachchi C, Boakye HT, Uden PC, Tyson JF. Chromatographic speciation of anionic and neutral selenium compounds in Se-accumulating *Brassica juncea* (Indian mustard) and selenized yeast. Journal of Chromatography. A. 2004;**1054**:303-312. DOI: 10.1016/j.chroma.2004.07.083

[51] Mounicou S, McSheehy S, Szpunar J, Potin-Gautier M, Łobiński R. Analysis of selenium yeast for selenium speciation by size-exclusion chromatography and capillary zone electrophoresis with inductively coupled plasma mass spectrometric detection (SEC-CZE-ICP-MS). Journal of Analytical Atomic Spectrometry. 2002;**17**:15-20. DOI: 10.1039/B107943N

[52] Chery CC, Gunther D, Cornelis R, Vanhaecke F, Moens L. Detection of metals in proteins by means of polyacrylamide gel electrophoresis and laser ablation-inductively coupled plasma-mass spectrometry: Application to selenium. Electrophoresis. 2003;**24**:3305-3313. DOI: 10.1002/elps.200305590

[53] Ayouni L. Speciation of selenium in a commercial dietary supplement by liquid chromatography coupled with inductively coupled plasma-mass spectrometry (ICP-MS). Journal of Toxicology and Environmental Health, Part A. 2007;**70**:735-741. DOI: 10.1080/15287390701236314

[54] Moreno P, Quijano MA, Gutierrez AM, Perez-Conde MC, Camara C. Study of selenium species distribution in biological tissues by size exclusion and ion exchange chromatography inductively coupled plasma-mass spectrometry. Analytica Chimica Acta. 2004;**524**:315-327. DOI: 10.1016/j.aca.2004.02.029

[55] Siwek M, Galunsky B, Niemeyer B. Isolation of selenium organic species from arctic krill after enzymatic hydrolysis. Analytical and Bioanalytical Chemistry. 2005;**281**:737-741. DOI: 10.1007/s00216-004-2936-z

[56] Koyama H, Omura K, Ejima A, Kasanuma Y, Watanabe C, Satoh H. Separation of selenium-containing proteins in human and mouse plasma using tandem high-performance liquid chromatography columns coupled with inductively coupled plasma mass spectrometry. Analytical Biochemistry. 1999;**267**:84-91. DOI: 10.1006/abio.1998.2949

[57] Kobayasi Y, Orga Y, Suzuki KT. Speciation and metabolism of selenium injected with 82Se-enriched selenite and selenite in rats. Journal of Chromatography B: Biomedical Sciences and Applications. 2001;**760**:73-81. DOI: 10.1016/S0378-4347(01)00252-3

[58] Duncan EG, Maher WA, Jagtap R, Krikowa F, Roper MM, O'Sullivan CA. Selenium speciation in wheat grain varies in the presence of nitrogen and sulphur fertilizers. Environmental Geochemistry and Health. 2017;**39**:955-966. DOI: 10.1007/s10653-016-9857-6

[59] Lavu RVS, Laing GD, Van De Wiele T, Pratti VL, Willekens K, Vandecasteele B, et al. Fertilizing soil with selenium fertilizers: Impact on concentration, speciation, and bioaccessibility of selenium in leek (*Allium ampeloprasum*). Journal of Agricultural and Food Chemistry. 2012;**60**: 10930-10935. DOI: 10.1021/jf302931z

[60] Garousi F, Domokos-Szabolesy F, Jánószky M, Kovács AB, Veres S, Soós A, et al. Selenoamino acid-enriched green pea as a value-added plant protein source for humans and livestock. Plant Foods for Human Nutrition. 2017;**73**:168-175. DOI: 10.1007/s11130-017-0606-5

[61] Ni Z, Liu Y, Qu M. Determination of 5 natural selenium species in selenium-enriched bamboo shoots using LC-ICP-MS. Food Science and Biotechnology. 2014;**23**:1049-1053. DOI: 10.1007/s10068-014-0143-z

[62] Gomes da Silva E, Mataveli LRV, Arruda MAZ. Speciation analysis of selenium in plankton, Brazil nut and human urine samples by HPLC-ICP-MS. Talanta. 2013;**110**:53-57. DOI: 10.1016/j.talanta.2013.02.014

[63] Michalska-Kacymirow M, Kurek E, Smolis A, Wierzbicka M, Bulska E. Biological and chemical investigation of *Allium cepa* L. response to selenium inorganic compounds. Analytical and Bioanalytical Chemistry. 2014;**406**:3717-3722. DOI: 10.1007/s00216-014-7742-7

[64] Ruszczyńska A, Konopka A, Kurek E, Elguera JCT, Bulska E. Investigation of biotransformation of selenium in plants using spectrometric methods. Spectrochimica Acta Part B. 2017;**130**:7-16. DOI: 10.1016/j.sab.2017.02.004

[65] Larsen EH, Sloth J, Hansen M, Moesgaard S. Selenium speciation and isotope composition in 77Se enriched yeast using gradient elution HPLC separation and ICP-dynamic reaction cell-MS. Journal of Analytical Atomic Spectrometry. 2003;**18**:310-316. DOI: 10.1039/B209586F

[66] Khanam Am Platel K. Bioaccessibility of selenium, selenomethionine and selenocysteine from foods and influence of heat processing on the same. Food Chemistry. 2016;**184**:1293-1299. DOI: 10.1016/j.foodchem.2015.09.005

[67] Zhang HJ, Gao PF, Guo XF, Wang H. Simultaneous determination of selenium containing amino acids and their sulfur-analogues in green tea and gynostemma pentaphyllum infusion with high performance liquid chromatography based on fluorescence labeling. Microchemical Journal. 2013;**110**:192-197. DOI: 10.1016/j.microc.2013.03.016

[68] Tie M, Li B, Zhuang X, Han J, Liu L, Hu Y, et al. Selenium speciation in soybean by high performance liquid chromatography coupled to electrospray ionization-tandem mass spectrometry (HPLC-ESI-MS/MS). Microchemical Journal. 2015;**123**:70-75. DOI: 10.1016/j.microc.2015.05.017

[69] Maneetong S, Chookhampaeng S, Chantiratikul A, Chinrasri O, Thosaikham W, Sittipout R, et al. Hydroponic cultivation of selenium-enriched kale (*Brassica oleracea var. alboglabra* L.) seedling and speciation of selenium with HPLC-ICP-MS. Microchemical Journal. 2013;**108**:87-91. DOI: 10.1016/j.microc.2013.01.003

[70] Vonderheide AP, Wrobel K, Kannamkumarath SS, B'Hymer C, Montes-Bayon M, de Leon CP, et al. Characterization of selenium species in Brazil nuts by HPLC-ICP-MS and ES-MS. Journal of Agricultural and Food Chemistry. 2002;**50**:5722-5728. DOI: 10.1021/jf0256541

[71] Marchante-Gayon JM, Thomas C, Feldmann I, Jakubowski N. Comparison of different nebulizers and chromatographic techniques for the separation of selenium in nutritional commercial supplements by hexapole collision and reaction cell ICP-MS. Journal of Analytical Atomic Spectrometry. 2000;**15**:1093-1102. DOI: 10.1039/B002372H

[72] Zheng J, Kosmus W. Retention study of inorganic and organic selenium compounds on a silica-based reversed phase column with ion-pairing reagents. Chromatographia. 2000;**51**:338-344. DOI: 10.1007/BF02490613

[73] Garcia-Alvarez-Coque MC, Ruiz-Angel MJ, Berthod A, Carda-Broch S. On the use of ionic liquids as mobile phase additives in high-performance liquid chromatography. Review Analytica Chimica Acta. 2015;**883**:1-21. DOI: 10.1016/j.aca.2015.03.042

[74] Grijalba AC, Fiorentini EF, Wuilloud RG. Ionic liquid-assisted separation and determination of selenium species in food and beverage samples by liquid chromatography coupled to hydride generation atomic fluorescence spectrometry. Journal of Chromatography. A. 2017;**1491**:227-125. DOI: 10.1016/j.chroma.2017.02.045

[75] McCalley DV. Understanding and manipulating the separation in hydrophilic interaction liquid chromatography—A review. Journal of

Chromatography. A. 2017;**1523**:49-71. DOI: 10.1016/j.chroma.2017.06.026

[76] Far J, Preud'homme H, Lobinski R. Detection and identification of hydrophilic selenium compounds in selenium-rich yeast by size exclusion-microbore normal-phase HPLC with the on-line ICP-MS and electrospray Q-TOF-MS detection. Analytica Chimica Acta. 2010;**657**:175-190. DOI: 10.1016/j.aca.2009.10.040

[77] Dernovics M, Far J, Lobinski R. Identification of anionic selenium species in Se-rich yeast by electrospray QTOF MS/MS and hybrid linear ion trap/orbitrap MSn. Metallomics. 2009;**1**:317-329. DOI: 10.1039/b901184f

[78] Greco G, Letzel T. Main interactions and influences of the chromatographic parameters in HILIC separation. Journal of Chromatographic Science. 2013;**51**:684-693. DOI: 10.1093/chromsci/bmt015

[79] Sentkowska A, Pyrzynska K. Hydrophilic interaction liquid chromatography in the speciation analysis of selenium. Journal of Chromatography B. 2018;**1074-1075**:8-15. DOI: 10.1016/jchromb.2018.01.006

[80] Arnaudguilhem C, Bierla K, Ouerdane L, Preud'homme H, Yiannikouris Y, Lobinski R. Selenium metabolomics in yeast using complementary reversed-phase/hydrophilic ion interaction (HILIC) liquid chromatography-electrospray hybrid quadrupole trap/orbitrap mass spectrometry. Analytica Chimica Acta. 2012;**757**:26-38. DOI: 10.1016/j.aca.2012.10.029

[81] Bierła K, Godin S, Lobiński R, Szpunar J. Advances in electrospray mass spectrometry for the selenium speciation: Focus on Se-rich yeast. TRAC Trends in Analytical Chemistry. 2018;**104**:87-94

Chapter 6

Selenium in the Prevention and Treatment of Hepatocellular Carcinoma: From Biomedical Investigation to Clinical Application

Chien-Shan Cheng, Ning Wang and Yibin Feng

Abstract

Selenium is a micronutrient that had been suggested to reduce the risk of cancer. Hepatocellular carcinoma (HCC), a prevalent disease and one of the most lethal cancers in the world, awaits new alternative treatment strategies to improve patients' survival. As an essential trace element, selenium has been studied for its anticancer properties in both oxidative stress and inflammatory-related mechanisms that may contribute to HCC growth and metastasis. In recent decades, increasing studies have investigated the potential role of selenium in liver cancer involving several major cancer-associated signaling pathways, metabolic pathways, and antioxidant defense systems both *in vitro* and in preclinical models. It was also observed that there was an increase in the trend of development of novel selenium nanoparticles and selenium-containing inhibitors aiming to improve the therapeutic efficacy and relative potency of selenium. However, controversies remain with whether a relationship exists between serum selenium level and HCC risk. This chapter aims to summarize the multi-target and multi-pathway *in vitro* and *in vivo* pharmacological effects of selenium in HCC, to provide a more comprehensive view and to highlight the recently discovered molecular mechanisms We hope this chapter could outline the correlation of selenium level and the risk of HCC in patients and discuss the clinical application of selenium in HCC prevention and treatment.

Keywords: selenium, hepatocellular carcinoma, anti-cancer, carcinogenesis

1. Introduction

Selenium (Se) is a naturally occurring essential micronutrient which had been received considerable attention in medicine and biology. As an essential component for mammalian cellular function, for the synthesis of several antioxidant enzymes, such as glutathione peroxidase, and for the synthesis of selenoproteins, selenium not only plays a vital role in balancing the redox environment in the body, but also is related to the prevention of diseases related to free radical damage, including infectious diseases, cancer, and cardiovascular diseases [1, 2]. The relationship between selenium and health has been the focus of medical community [3]. While early observational studies have shown that the trace element selenium in the environment is closely related to the occurrence and development of tumors [3]. The incidence of cancer in patients with selenium deficiency is significantly increased, and the amount of selenium in the body is negatively correlated with cancer [4]. Furthermore, while some studies suggested selenium supplementation reduces the risk of cancer, some methodologically sound trials suggested selenium supplementation does not reduce the risk of cancer and may even increase it for some types, including advanced prostate cancer and skin cancer [5, 6]. An increased risk of diabetes has also been reported [7]. The relationship between selenium and cancer has been one of the most heated debates in human health over the past few decades. Moreover, the potential effect of selenium in the prevention and treatment of HCC is worthy of further investigation.

Primary liver cancer (PLC) is one of the most prevalent malignancies worldwide, and its incidence ranks fourth among all malignancies in China [8, 9]. The overall prognosis for patients with PLC is unsatisfactory with a 5-year survival rate of about 18%, and its mortality rate ranks the third in tumor-related death [10]. HCC is the most prevalent pathological type of PLC, accounting for about 85–90% of all PLCs [11]. Among all HCC patients, approximately 70–90% of them presented with a history of chronic liver diseases and cirrhosis [12]. While the primary risk factors for developing cirrhosis include chronic hepatitis B virus (HBV), hepatitis C virus (HCV), alcoholic liver disease, and nonalcoholic steatohepatitis (NASH), some other risk factors for HCC include uptake of aflatoxin-contaminated food, diabetes, obesity, certain genetic diseases such as hemochromatosis and some metabolic disorders [9, 11, 13]. Also, geographical factors have a direct impact on the various etiological features of HCC patients and make HCC an extremely complex condition associated with poor prognosis.

Epidemiology studies have found that the geographical factors and specific environmental exposures may be associated with HCC [14]. Inconsistent with the world's selenium distribution patterns, previous epidemiological studies have highlighted the particular relevance of selenium-associated cancer risk to the region of low selenium distribution, such as Asia or Africa [5, 15]. Some earlier surveying studies found that low environmental selenium is associated with certain cancers in the digestive system, and selenium supplementation may provide some cancer prevention effect [16]. However, the results had been inconsistent throughout various studies. Previous intervention trials in regions with low environmental selenium have shown a beneficial effect of selenium supplementation for cancer patients, while in most parts of North America, where there is sufficient environmental selenium, supplemental intake of selenium appears to be unrelated to chemoprevention [5, 17].

In Qidong, Jiangsu province, China, there is a high incidence of HCC and low environmental selenium level [18]. Various studies carried out in the region unraveled the negative correlation of HCC mortality with serum selenium level and environmental selenium level [19, 20]. However, in other areas in the world with low environmental selenium level, such correlation was not observed, suggesting that selenium deficiency may not be the sole cause of liver cancer and other factors, such as environmental carcinogens, the of heavy industry, chemical industry, etc. may collaboratively contribute to the high incidence of HCC [4].

Since the 1970s, continuous efforts had been made on studying the potential anti-cancer effect of selenium supplementation. Decrease of serum selenium level has also been observed in patients with chronic liver diseases. Mechanically, selenium plays a vital role in maintaining healthy liver function and synthesis of essential liver enzymes [21, 22]. Selenium protects the liver mainly in the following aspects: (1) reduces the damage of toxic substances in the environment to the liver;

DOI: http://dx.doi.org/10.5772/intechopen.88960

(2) protects the integrity of the liver cell membrane by scavenging free radicals and preventing lipid peroxides through glutathione peroxidase; (3) accelerates the catabolism of ethanol, protecting the liver from alcoholic damage; and (4) stimulates both the humoral and cellular immunity and enhances immune function against hepatitis virus. The present article reviews the roles of selenium in HCC.

2. The preventive role of selenium in HCC tumorigenesis

One of the most recognized roles of selenium is to prevent cancer. Since the 1970s, the tumorigenesis prevention role of selenium has also been studied in various chemical carcinogen-induced HCC models; for example, selenium can significantly suppress liver cancer carcinogenesis induced by aflatoxin B1, dimethyl azobenzene, or acetylamino. Consumption of food with Aflatoxin B1 (AFB1) contamination is one of the leading causes of liver cancer. In 1985, Milks, et al. investigated the effect of selenium on aflatoxin hepatocarcinogenesis in the rat model and found that oral supplementation of selenium can dose- and time-dependently reduces aflatoxin B1-induced γ-GT variant hepatocyte foci and tumor nodules in rat liver [23]. Similarly, another animal study by Lei et al. in 1990 also found that selenium supplementation had an inhibitory effect on the initiation and promotion stages of AFB1-induced preneoplastic foci and nodules in HCC model without evidence of toxicity [24]. Further study showed that oral supplementation of selenium could dose- and time-dependently reduces aflatoxin B1-induced γ-GT variant hepatocyte foci and tumor nodules in animal models. Shi et al. demonstrated that selenium could reduce the amount of AFB1-DNA binding and effectively inhibit AFB1-induced DNA damage [25]. The main reason may be that AFB1 binds to glutathione (GSH) under the catalysis of the glutathione-S-converting enzyme (GSTS) and is excreted in a non-toxic form, thereby reducing the formation of AFB1-DNA compounds.

Diethylnitrosamine (DENA)-induced and phenobarbital (PB)-promoted HCC model is one of the most popular chemically induced HCC models used [26]. Thirunavukkarasu et al. conducted various studies investigating the chemopreventive properties and biochemical mechanisms of sodium selenite supplementation in the DENA-initiated and PB-promoted HCC rodent model with four parts per million (p.p.m.) of sodium selenite in drinking water for 14 weeks. Selenium supplementation elevates malate dehydrogenase level, a vital enzyme of the citric acid cycle [27]; increases superoxide dismutase (SOD and catalase (CAT) levels, two key antioxidant enzymes [28, 29]; decreases glutathione transferase (GST) level [30]; decreases alanine transaminase (AST), aspartate transaminase (ALT), and lactate dehydrogenase (LDH) elevations [31]; suppresses the elevated glycoprotein levels of glycoproteins such as globulin and hexosamine [32]; elevates ATPase enzymatic level and alters serum mineral levels [33, 34]; as well as increases vitamins C and E [35] in the DENA-initiated and PB-promoted HCC rodent model.

A more recent study by Liu et al. investigated the effects of selenium-enriched malt (SEM) on hepatocarcinogenesis, paraneoplastic syndrome, the hormones regulating blood glucose, vascular endothelial growth factor (VEGF), and several relevant angiogenic cytokines in DENA-induced HCC rat model [36–38]. The results showed that SEM decreased several liver enzyme levels, including alanine aminotransferase (ALT), alkaline phosphatase (ALP), as well as gamma-glutamyltranspeptidase (GGT); increased glucose levels and reduced hypoglycemia; and inhibited the angiogenesis by downregulating the expression of VEGF and interacted with tumor necrosis factor-alpha (TNF-α) and insulin-like growth factor II (IGF-II), and nitric oxide (NO). These results suggested that SEM may delay the development of hepatocarcinoma in DENA-induced HCC rat model and partially by the inhibition of angiogenesis [38].

Other preclinical animal models had also been used to study the effect of selenium in liver cancer. Nine or thirteen weeks of sodium selenite (6 p.p.m) supplementation in drinking water prevented the azo dye (3′-MeDAB) hepatocarcinogenesis in male Sprague-Dawley rats [39, 40]. Supplementation with sodium selenite or selenomethionine reduces the focal volume and lower GST level in 2-acetylaminofluorene (2-AAF)-induced hepatocarcinogenesis rat model [41–43]. Sodium selenite suppressed the BOP-induced HCC in Syrian hamsters and induced apoptosis and caspase-3 level [44]. In the transgenic mice model, selenium supplementation also demonstrates anti-hepatocarcinogenesis effect. For example, in CBA mice, prenatal administration of selenium significantly decreased the spontaneous liver tumorigenesis [45]; in TGF-α/c-Myc transgenic mice, dietary selenium supplementation inhibited hepatocarcinogenesis, promoted apoptosis, and suppressed cell proliferation [46]; in Mdr2 knock-out mice, selenomethionine suppressed the development of HCC by regulating various genes involved in inflammation and oxidative stress [47]. In HCC xenograft, selenium-enriched green tea extracts also showed a suppressive effect in human hepatoma HepG2 cell mice xenograft model [48]. These results suggested that selenium holds a significant promising role as a potential anti-cancer agent *in vivo* in various liver cancer models.

3. Anti-cancer effect of selenium and the potential underlying mechanism

Since the 1970s, the potential anti-tumor effects of selenium have attracted considerable research attention. In vitro studies have shown proliferation inhibitory role of selenium in liver cancer, colon cancer, gastric cancer, etc. [49, 50]. Various *in vivo* and *in vitro* studies have revealed its promising anti-tumor role breast cancer, lung cancer, colorectal cancer, etc. [51, 52]. These results suggest the broad-spectrum anti-cancer effect of selenium. In recent decades, increasing studies have investigated the potential role of selenium in liver cancer, involving several major cancer-associated signaling pathways, metabolic pathways, and transcription factors.

4. Anti-proliferative and apoptosis induction role of selenium in HCC

One of the earlier studies by Baker et al. in the 1990s on human hepatoma Hep3B cell line showed that selenium supplementation restores glutathione peroxidase mRNA expression in selenium-deprived cell culture [53]. Hill et al. investigated in human liver cancer HepG2 cell line showed that in selenium-deprived HepG2 cells, selenoprotein P release decreased to 10% [54]. Further, various studies consistently reported apoptosis induction effect of selenite in human hepatoma cells HepG2 cells, potentially by inducing the release of lactate dehydrogenase (LDH) and decreasing glutathione (GSH) production [55–57]. Another study reported that selenite-induced apoptosis in HepG2 cells was mediated by reactive oxygen species (ROS) that activated JNK to regulate apoptosis [58]. A more recent study on selenium nanoparticle surface decorated with galangin can induce apoptosis through p38 and AKT signaling pathway in HepG2 cells [59]. Similarly, selenium nanoparticles synthesized with extract of hawthorn fruit also induced apoptosis in HepG2 cells [60]. Various mechanisms may be involved in the apoptosis induction role of selenium and selenium-containing inhibitors for HCC. Tagaram et al. reported a selenium-containing MAPK and PI3 kinase inhibitor, the *Se*, *Se'*-1,4-phenylenebis(1,2-ethanediyl) bisisoselenourea (PBISe), possesses anti-proliferative and pro-apoptotic ability in HCC cell lines *in vitro* and in *in vivo* in a spontaneous murine HCC model [39]. PBISe promoted apoptosis by inhibiting PI3K, MAPK, and STAT3 signaling *in vitro* and with significant reduce tumor sizes *in vivo* ($p < 0.007$) with a significant reduction in tumor survival marker PCNA and angiogenesis markers Vegf-A, Vegf-R3, and CD34 [39]. These results demonstrate the chemotherapeutic effects of selenium by inhibiting tumor proliferation and inducing tumor apoptosis for HCC treatment.

5. Anti-oxidation effect of selenium in HCC

Oxidative stress is characterized by the excessive production of oxidants or ROS and insufficient elimination by protective antioxidants [40]. This persistent imbalance may lead to somatic mutations and neoplastic transformation [40]. In cancer, ROS is commonly found elevated and is a key constituent to cancer survival [61, 62]. Most of the effects of dietary selenium on oxidation are attributable to the insertion of this element to selenoproteins, mostly its cofactor glutathione peroxidase. In the active oxidative metabolism, selenium-dependent glutathione peroxidase acts with tripeptide glutathione (GSH) and competes for the catalyzation for hydrogen peroxide as a substrate, scavenging reactive oxygen species (ROS) and lipid peroxide (LPO) [63]. In 2007, a study by Katzenellenbogen et al. investigated the effect of selenomethionine on oxidative stress and inflammation or lipid metabolism with cDNA microarrays in HCC development in Mdr2 knockout mice model [47]. The results showed that selenomethionine alters the expression of commonly upregulated genes found in response to inflammation, oxidative stress, and cancer [47]. Further, the inhibitory effect on gene expression of selenomethionine is positively correlated with its role to reduce the incidence of large tumors [47]. These results provide a theoretical basis for the anticancer mechanism of selenium.

6. Anti-metastasis effect of selenium in HCC

HCC is characterized by its invasive and metastatic potential [64]. In liver cancer, it has been suggested that SBP1, a Se-containing protein, and its primary function is Se transport, plays a role in metastasis and if found to be highly expressed in healthy liver tissue but was nearly non-detectable in highly metastatic liver cancer cell lines [65]. Epithelial-mesenchymal transition (EMT) is a process that plays a vital role in HCC metastasis cascade and had been suggested to be closely related to the initiation of HIF-1α [64]. SBP-1 is a downstream target of HIF-1α and has been found that loss of SBP1 promoted liver cancer cell migration and increased GPX1 activity, which further suppresses in hydrogen peroxide and other reactive oxygen species, leading to the inhibition of HIF-1α [64]. Recently study by Gao et al. identified 186 differentially expressed genes among control and SBP1 expressing HCC cells [66]. Further investigation showed C-X-C motif chemokine receptor 4 (CXCR4) expression was inhibited by SBP1 and is closely related to the migration and invasion ability of HCC cells through activation of AKT signaling [66]. These results suggested the potential application of selenium in liver cancer metastasis prevention and treatment; however, more data from *in vitro* and *in vivo* are warranted to solidify effect and mechanism of action.

7. Epidemiological investigations and clinical trials

Many studies have shown that the level of blood selenium in patients with chronic liver disease is reduced, and it can be corrected by selenium supplementation [67–69]. Selenium supplementation can inhibit the lipid peroxidation of immune cells in patients with liver diseases, regulate cellular immune function, and alleviate immunopathological damage. In the treatment of patients with chronic hepatitis, in the anti-viral, immune-modulating, and liver-protecting treatment, it is beneficial to correct the blood selenium level of hepatitis patients by detecting and conditioning so that lipid oxidation and liver repair could be resisted [12]. Selenium and selenium supplementation for the treatment of liver disease should attract the attention of the medical community. However, controversies remain with whether a relationship exists between serum selenium level and HCC risk.

Some pioneering studies performed in China in the 1990s showed that plasma selenium level is negatively correlated with the occurrence of HCC in areas with low environmental selenium level [19, 20]. A preliminary report by Yu et al. in 1991 of two intervention trials in high-risk populations of PLC with selenium supplementation in Qidong, Jiangsu province, China, showed that among individuals with a family history of PLC, oral supplementation of 200 mg selenium in the form of selenized yeast (Se-yeast) daily for 4 and 2 years, respectively, showed significant reduction in PLC incidence compared with placebo [19]. Another study by Yu et al. in 1997 of an interventional trial among 130,471 individuals living in Qidong Country showed that a reduction PLC incidence by 35.1% in selenite table salt supplemented group vs. the non-supplemented group after an 8-year follow-up. Consistently, an epidemiological study in Taiwan by Yu et al. involving a cohort of 7342 chronic carriers of hepatitis B and C virus male with an average of 5.3 years of follow-up showed an inverse association between plasma selenium levels and the risk of HCC among men with chronic hepatitis virus infection [70]. Another study in Korea by Kim et al. in 2012 also reported an apparent correlation between low plasma selenium level and incidence of HCC in Korean hepatoma patients [71]. However, no relationship was observed between plasma selenium concentration and incidence of HCC in a study with Japanese hepatoma patients [72]. In 2016, a systemic review meta-analysis involving 9 trials and a total of 1433 subjects supported an inverse correlation between Se level and the risk of HCC, and lowered Se level had a relationship with HCC with a remarkable heterogeneity ($I2 = 74.3\%$, $P < 0.001$) [4]. However, epidemiological investigations and biological studies should be further conducted to demonstrate and verify whether selenium supplements are beneficial for the prevention and treatment of HCC and to elucidate its exact mechanism of action.

8. Discussion and conclusion

Over decades, an increase in the trend of development of novel selenium nanoparticles and selenium-containing inhibitors aims to improve the therapeutic efficacy and relative potency of selenium. Experimental studies with animal tumor models and epidemiological studies of human tumors have revealed that selenium is one of the factors affecting the risk of cancer. The huge number of publications has suggested the potential role of selenium and redox-active selenium compounds as inhibitors and therapeutic agents for liver cancer. However, the studies are difficult to compare among different selenium compounds due to a high degree of variations on the effective dosage anti-proliferation due to the difference in the *ex vivo* culture condition.

Further, although some epidemiology studies have supported the hypothesis that selenium supplementation can reduce the risk of cancer, inconsistent results have also been reported. Another important aspect of being considered for the use of selenium-containing compounds in anticancer is as a chemical protective agent against toxic side effects of anticancer drugs. Selenium has also been reported to have a protective effect against cisplatin-induced nephrotoxicity without affecting its anti-tumor activity in rodent model [73]. Further trials are needed to confirm the selenium supplementation for liver cancer in terms of its effect on prognosis as well as potential toxic effect.

This chapter summarizes the pharmacological effects of selenium in HCC *in vitro* and *in vivo*, and to highlight the recently discovered molecular mechanisms, to outline the correlation of selenium level and the risk of HCC in patients, and to discuss the clinical application of selenium in HCC prevention and treatment. In conclusion, the preclinical studies presented in this chapter summarized the promise of selenium as a potential anti-cancer agent and presented the chemo-preventive role in HCC. Although substantial evidence for the anti-cancer effect of selenium is discussed, further studies are warranted.

Author details

Chien-Shan Cheng, Ning Wang and Yibin Feng*
School of Chinese Medicine, The University of Hong Kong, Hong Kong

*Address all correspondence to: yfeng@hku.hk

References

[1] Schomburg L, Schweizer U, Kohrle J. Selenium and selenoproteins in mammals: Extraordinary, essential, enigmatic. Cellular and Molecular Life Sciences. 2004;**61**(16):1988-1995

[2] Arigony AL, de Oliveira IM, Machado M, et al. The influence of micronutrients in cell culture: A reflection on viability and genomic stability. BioMed Research International. 2013;**2013**:597282

[3] Vinceti M, Filippini T, Cilloni S, et al. Health risk assessment of environmental selenium: Emerging evidence and challenges (Review). Molecular Medicine Reports. 2017;**15**(5):3323-3335

[4] Zhang Z, Bi M, Liu Q, Yang J, Xu S. Meta-analysis of the correlation between selenium and incidence of hepatocellular carcinoma. Oncotarget. 2016;**7**(47):77110-77116

[5] Vinceti M, Filippini T, Del Giovane C, et al. Selenium for preventing cancer. Cochrane Database of Systematic Reviews. 2018;**1**:CD005195

[6] Vinceti M, Filippini T, Cilloni S, Crespi CM. The epidemiology of selenium and human cancer. Advances in Cancer Research. 2017;**136**:1-48

[7] Koyama H, Mutakin, Abdulah R, Yamazaki C, Kameo S. Selenium supplementation trials for cancer prevention and the subsequent risk of type 2 diabetes mellitus: Selenium and vitamin E cancer prevention trial and after. Nihon Eiseigaku Zasshi. 2013;**68**(1):1-10

[8] Zhang Y, Ren JS, Shi JF, et al. International trends in primary liver cancer incidence from 1973 to 2007. BMC Cancer. 2015;**15**:94

[9] Ghouri YA, Mian I, Rowe JH. Review of hepatocellular carcinoma: Epidemiology, etiology, and carcinogenesis. Journal of Carcinogenesis. 2017;**16**:1

[10] Balogh J, Victor D 3rd, Asham EH, et al. Hepatocellular carcinoma: A review. Journal of Hepatocellular Carcinoma. 2016;**3**:41-53

[11] Nordenstedt H, White DL, El-Serag HB. The changing pattern of epidemiology in hepatocellular carcinoma. Digestive and Liver Disease. 2010;**42**(Suppl 3):S206-S214

[12] Ascha MS, Hanouneh IA, Lopez R, Tamimi TA, Feldstein AF, Zein NN. The incidence and risk factors of hepatocellular carcinoma in patients with nonalcoholic steatohepatitis. Hepatology. 2010;**51**(6):1972-1978

[13] Thuluvath PJ, Triger DR. Selenium in chronic liver disease. Journal of Hepatology. 1992;**14**(2-3):176-182

[14] Santella RM, Wu HC. Environmental exposures and hepatocellular carcinoma. Journal of Clinical and Translational Hepatology. 2013;**1**(2):138-143

[15] Vinceti M, Dennert G, Crespi CM, et al. Selenium for preventing cancer. Cochrane Database of Systematic Reviews. 2014;**2014**(3):CD005195

[16] Shenkin A. Selenium in intravenous nutrition. Gastroenterology. 2009; **137**(Suppl 5):S61-S69

[17] Lippman SM, Klein EA, Goodman PJ, et al. Effect of selenium and vitamin E on risk of prostate cancer and other cancers: The selenium and vitamin E cancer prevention trial (SELECT). Journal of the American Medical Association. 2009;**301**(1):39-51

[18] Sakoda LC, Graubard BI, Evans AA, et al. Toenail selenium

and risk of hepatocellular carcinoma mortality in Haimen City, China. International Journal of Cancer. 2005;**115**(4):618-624

[19] Yu SY, Zhu YJ, Li WG, et al. A preliminary report on the intervention trials of primary liver cancer in high-risk populations with nutritional supplementation of selenium in China. Biological Trace Element Research. 1991;**29**(3):289-294

[20] Yu SY, Zhu YJ, Li WG. Protective role of selenium against hepatitis B virus and primary liver cancer in Qidong. Biological Trace Element Research. 1997;**56**(1):117-124

[21] Gupta S, Read SA, Shackel NA, Hebbard L, George J, Ahlenstiel G. The role of micronutrients in the infection and subsequent response to hepatitis C virus. Cell. 2019;**8**(6):603

[22] Guerra TS, Hoehr NF, Boin Ide F, Stucchi RS. Trace elements in plasma and nutritional assessment in patients with compensated cirrhosis on a liver transplant list. Arquivos de Gastroenterologia. 2016;**53**(2):84-88

[23] Milks MM, Wilt SR, Ali II, Couri D. The effects of selenium on the emergence of aflatoxin B1-induced enzyme-altered foci in rat liver. Fundamental and Applied Toxicology. 1985;**5**(2):320-326

[24] Lei DN, Wang LQ, Ruebner BH, et al. Effect of selenium on aflatoxin hepatocarcinogenesis in the rat. Biomedical and Environmental Sciences. 1990;**3**(1):65-80

[25] Shi CY, Chua SC, Lee HP, Ong CN. Inhibition of aflatoxin B1-DNA binding and adduct formation by selenium in rats. Cancer Letters. 1994;**82**(2):203-208

[26] Bishayee A, Chatterjee M. Inhibitory effect of vanadium on rat liver carcinogenesis initiated with diethylnitrosamine and promoted by phenobarbital. British Journal of Cancer. 1995;**71**(6):1214-1220

[27] Thirunavukkarasu C, Singh JP, Selvendiran K, Sakthisekaran D. Chemopreventive efficacy of selenium against N-nitrosodiethylamine-induced hepatoma in albino rats. Cell Biochemistry and Function. 2001;**19**(4):265-271

[28] Thirunavukkarasu C, Sakthisekaran D. Effect of selenium on N-nitrosodiethylamine-induced multistage hepatocarcinogenesis with reference to lipid peroxidation and enzymic antioxidants. Cell Biochemistry and Function. 2001;**19**(1):27-35

[29] Thirunavukkarasu C, Prince Vijeya Singh J, Thangavel M, Selvendiran K, Sakthisekaran D. Dietary influence of selenium on the incidence of N-nitrosodiethylamine-induced hepatoma with reference to drug and glutathione metabolizing enzymes. Cell Biochemistry and Function. 2002;**20**(4):347-356

[30] Thirunavukkarasu C, Sakthisekaran D. Sodium selenite, dietary micronutrient, prevents the lymphocyte DNA damage induced by N-nitrosodiethylamine and phenobarbital promoted experimental hepatocarcinogenesis. Journal of Cellular Biochemistry. 2003;**88**(3):578-588

[31] Thirunavukkarasu C, Sakthisekaran D. Sodium selenite modulates tumour marker indices in N-nitrosodiethylamine-initiated and phenobarbital-promoted rat liver carcinogenesis. Cell Biochemistry and Function. 2003;**21**(2):147-153

[32] Thirunavukkarasu C, Sakthisekaran D. Influence of sodium selenite on glycoprotein contents in normal and N-nitrosodiethylamine

initiated and phenobarbital promoted rat liver tumors. Pharmacological Research. 2003;**48**(2):167-173

[33] Thirunavukkarasu C, Sakthisekaran D. Stabilization of membrane bound enzyme profiles by sodium selenite in N-nitrosodiethylamine induced and phenobarbital promoted hepatocarcinogenesis in rats. Biomedicine & Pharmacotherapy. 2003;**57**(3-4):117-123

[34] Thirunavukkarasu C, Sakthisekaran D. Effect of dietary selenite on N-nitrosodiethylamine-induced and phenobarbital promoted multistage hepatocarcinogenesis in rat: Reflection in some minerals. Biomedicine & Pharmacotherapy. 2003;**57**(9):416-421

[35] Thirunavukkarasu C, Babu E, Ebrahim AS, Chandramohan N, Sakthisekaran D. Antioxidant-associated chemoprevention by sodium selenite in N-nitrosodiethylamine-induced and phenobarbital-promoted hepatocarcinogenesis in rats. Cell Biochemistry and Function. 2004;**22**(4):265-271

[36] Liu JG, Zhao HJ, Liu YJ, Wang XL. Effect of selenium-enriched malt on hepatocarcinogenesis, paraneoplastic syndrome and the hormones regulating blood glucose in rats treated by diethylnitrosamine. Life Sciences. 2006;**78**(20):2315-2321

[37] Liu JG, Zhao HJ, Liu YJ, Wang XL. Effect of selenium-enriched malt on hypoglycemia and regulatory hormones in diethylnitrosamine-induced hepatocarcinoma SD rats. Research in Veterinary Science. 2009;**87**(3):438-444

[38] Liu JG, Zhao HJ, Liu YJ, Wang XL. Effect of selenium-enriched malt on VEGF and several relevant angiogenic cytokines in diethylnitrosamine-induced hepatocarcinoma rats. Journal of Trace Elements in Medicine and Biology. 2010;**24**(1):52-57

[39] Tagaram HR, Desai D, Li G, et al. A selenium containing inhibitor for the treatment of hepatocellular cancer. Pharmaceuticals. 2016;**9**(2):18

[40] Reuter S, Gupta SC, Chaturvedi MM, Aggarwal BB. Oxidative stress, inflammation, and cancer: How are they linked? Free Radical Biology & Medicine. 2010;**49**(11):1603-1616

[41] LeBoeuf RA, Laishes BA, Hoekstra WG. Effects of dietary selenium concentration on the development of enzyme-altered liver foci and hepatocellular carcinoma induced by diethylnitrosamine or N-acetylaminofluorene in rats. Cancer Research. 1985;**45**(11 Pt 1):5489-5495

[42] Liu Y, Yin T, Feng Y, et al. Mammalian models of chemically induced primary malignancies exploitable for imaging-based preclinical theragnostic research. Quantitative Imaging in Medicine and Surgery. 2015;**5**(5):708-729

[43] Mukherjee B, Ghosh S, Chatterjee M. Chemopreventive efficacy of selenomethionine and its role in the antioxidant defense system in 2-acetylaminofluorene-induced hepatocarcinogenesis in rats. Journal of Experimental Therapeutics & Oncology. 1996;**1**(4):209-217

[44] Lee CY, Hsu YC, Wang JY, Chen CC, Chiu JH. Chemopreventive effect of selenium and Chinese medicinal herbs on N-nitrosobis(2-oxopropyl)amine-induced hepatocellular carcinoma in Syrian hamsters. Liver International. 2008;**28**(6):841-855

[45] Popova NV. Perinatal selenium exposure decreases spontaneous liver tumorogenesis in CBA mice. Cancer Letters. 2002;**179**(1):39-42

[46] Novoselov SV, Calvisi DF, Labunskyy VM, et al. Selenoprotein deficiency and high levels of selenium compounds can effectively inhibit hepatocarcinogenesis in transgenic mice. Oncogene. 2005;**24**(54):8003-8011

[47] Katzenellenbogen M, Mizrahi L, Pappo O, et al. Molecular mechanisms of the chemopreventive effect on hepatocellular carcinoma development in Mdr2 knockout mice. Molecular Cancer Therapeutics. 2007;**6**(4):1283-1291

[48] Xu J, Yang F, An X, Hu Q. Anticarcinogenic activity of selenium-enriched green tea extracts in vivo. Journal of Agricultural and Food Chemistry. 2007;**55**(13):5349-5353

[49] Redman C, Scott JA, Baines AT, et al. Inhibitory effect of selenomethionine on the growth of three selected human tumor cell lines. Cancer Letters. 1998;**125**(1-2):103-110

[50] Chen YC, Prabhu KS, Mastro AM. Is selenium a potential treatment for cancer metastasis? Nutrients. 2013;**5**(4):1149-1168

[51] Peters U, Takata Y. Selenium and the prevention of prostate and colorectal cancer. Molecular Nutrition & Food Research. 2008;**52**(11):1261-1272

[52] Li D, Graef GL, Yee JA, Yan L. Dietary supplementation with high-selenium soy protein reduces pulmonary metastasis of melanoma cells in mice. The Journal of Nutrition. 2004;**134**(6):1536-1540

[53] Baker RD, Baker SS, LaRosa K, Whitney C, Newburger PE. Selenium regulation of glutathione peroxidase in human hepatoma cell line Hep3B. Archives of Biochemistry and Biophysics. 1993;**304**(1):53-57

[54] Hill KE, Chittum HS, Lyons PR, Boeglin ME, Burk RF. Effect of selenium on selenoprotein P expression in cultured liver cells. Biochimica et Biophysica Acta. 1996;**1313**(1):29-34

[55] Shen H, Yang C, Liu J, Ong C. Dual role of glutathione in selenite-induced oxidative stress and apoptosis in human hepatoma cells. Free Radical Biology & Medicine. 2000;**28**(7):1115-1124

[56] Shen HM, Ding WX, Ong CN. Intracellular glutathione is a cofactor in methylseleninic acid-induced apoptotic cell death of human hepatoma HEPG(2) cells. Free Radical Biology & Medicine. 2002;**33**(4):552-561

[57] Celik HA, Aydin HH, Deveci R, et al. Biochemical and morphological characteristics of selenite-induced apoptosis in human hepatoma hep G2 cells. Biological Trace Element Research. 2004;**99**(1-3):27-40

[58] Zou Y, Niu P, Yang J, Yuan J, Wu T, Chen X. The JNK signaling pathway is involved in sodium-selenite-induced apoptosis mediated by reactive oxygen in HepG2 cells. Cancer Biology & Therapy. 2008;**7**(5):689-696

[59] Li Y, Guo M, Lin Z, et al. Multifunctional selenium nanoparticles with Galangin-induced HepG2 cell apoptosis through p38 and AKT signalling pathway. Royal Society Open Science. 2018;**5**(11):180509

[60] Cui D, Liang T, Sun L, et al. Green synthesis of selenium nanoparticles with extract of hawthorn fruit induced HepG2 cells apoptosis. Pharmaceutical Biology. 2018;**56**(1):528-534

[61] Liou GY, Storz P. Reactive oxygen species in cancer. Free Radical Research. 2010;**44**(5):479-496

[62] Kumari S, Badana AK, G MM, G S, Malla R. Reactive oxygen species: A key constituent in cancer survival. Biomarker Insights. 2018;**13**:1177271918755391

[63] Ekoue DN, He C, Diamond AM, Bonini MG. Manganese superoxide dismutase and glutathione peroxidase-1 contribute to the rise and fall of mitochondrial reactive oxygen species which drive oncogenesis. Biochimica et Biophysica Acta—Bioenergetics. 2017;**1858**(8):628-632

[64] Christiansen JJ, Rajasekaran AK. Reassessing epithelial to mesenchymal transition as a prerequisite for carcinoma invasion and metastasis. Cancer Research. 2006;**66**(17):8319-8326

[65] Huang C, Ding G, Gu C, et al. Decreased selenium-binding protein 1 enhances glutathione peroxidase 1 activity and downregulates HIF-1alpha to promote hepatocellular carcinoma invasiveness. Clinical Cancer Research. 2012;**18**(11):3042-3053

[66] Gao PT, Ding GY, Yang X, et al. Invasive potential of hepatocellular carcinoma is enhanced by loss of selenium-binding protein 1 and subsequent upregulation of CXCR4. American Journal of Cancer Research. 2018;**8**(6):1040-1049

[67] Prystupa A, Kicinski P, Luchowska-Kocot D, et al. Association between serum selenium concentrations and levels of Proinflammatory and Profibrotic cytokines-Interleukin-6 and growth differentiation Factor-15, in patients with alcoholic liver cirrhosis. International Journal of Environmental Research and Public Health. 2017;**14**(4):437

[68] Shidfar F, Faghihi A, Amiri HL, Mousavi SN. Regression of nonalcoholic fatty liver disease with zinc and selenium Co-supplementation after disease progression in rats. Iranian Journal of Medical Sciences. 2018;**43**(1):26-31

[69] Burk RF, Hill KE, Motley AK, Byrne DW, Norsworthy BK. Selenium deficiency occurs in some patients with moderate-to-severe cirrhosis and can be corrected by administration of selenate but not selenomethionine: A randomized controlled trial. The American Journal of Clinical Nutrition. 2015;**102**(5):1126-1133

[70] Yu MW, Horng IS, Hsu KH, Chiang YC, Liaw YF, Chen CJ. Plasma selenium levels and risk of hepatocellular carcinoma among men with chronic hepatitis virus infection. American Journal of Epidemiology. 1999;**150**(4):367-374

[71] Kim IW, Bae SM, Kim YW, et al. Serum selenium levels in Korean hepatoma patients. Biological Trace Element Research. 2012;**148**(1):25-31

[72] Tashiro H, Kawamoto T, Okubo T, Koide O. Variation in the distribution of trace elements in hepatoma. Biological Trace Element Research. 2003;**95**(1):49-63

[73] Karavelioglu E, Boyaci MG, Simsek N, et al. Selenium protects cerebral cells by cisplatin induced neurotoxicity. Acta Cirúrgica Brasileira. 2015;**30**(6):394-400